T0205870

Reliability and Safety of Cable-Supported Bridges

Taylor and Francis Series in Resilience and Sustainability in Civil, Mechanical, Aerospace and Manufacturing Engineering Systems

Series Editor

Mohammad Noori
Cal Poly San Luis Obispo
Published Titles

Resilience of Critical Infrastructure Systems
Emerging Developments and Future Challenges
Zhishen Wu, Xilin Lu, Mohammad Noori

Experimental Vibration Analysis for Civil Structures
Testing, Sensing, Monitoring, and Control
Jian Zhang, Zhishen Wu, Mohammad Noori, and Yong Li

Reliability and Safety of Cable-Supported Bridges
Naiwei Lu, Yang Liu, and Mohammad Noori

Reliability-Based Analysis and Design of Structures and Infrastructure
Ehsan Noroozinejad Farsangi, Mohammad Noori, Paolo Gardoni,
Izuru Takewaki, Humberto Varum, Aleksandra Bogdanovic

For more information about this series, please visit: https://www.routledge.com/
Resilience-and-Sustainability-in-Civil-Mechanical-Aerospace-and-Manufacturing/
book-series/ENG

Reliability and Safety of Cable-Supported Bridges

Edited by

Naiwei Lu
Yang Liu
Mohammad Noori

CRC Press
Taylor & Francis Group
Boca Raton London New York

CRC Press is an imprint of the
Taylor & Francis Group, an **informa** business

First edition published 2021
by CRC Press
6000 Broken Sound Parkway NW, Suite 300, Boca Raton, FL 33487-2742

and by CRC Press
2 Park Square, Milton Park, Abingdon, Oxon, OX14 4RN

Library of Congress Cataloging-in-Publication Data
Names: Lu, Naiwei, editor. | Liu, Yang (Of Haerbin gong ye da xue), editor. | Noori, Mohammad, editor.
Title: Reliability and safety of cable-supported bridges / edited by Naiwei Lu, School of Civil Engineering, Changsha University of Science and Technology, Changsha, China, Yang Liu, School of Civil Engineering, Hunan University of Technology, Zhuzhou, China, Mohammad Noori, Dept. of Mechanical Engineering, California Polytechnic State Univ., San Luis Obispo, CA, United States.
Description: First edition. | Boca Raton : CRC Press, 2021. |
Series: Resilience and sustainability | Includes bibliographical references and index.
Identifiers: LCCN 2020052976 (print) | LCCN 2020052977 (ebook) |
  ISBN 9780367770266 (hbk) | ISBN 9781003170594 (ebk)
Subjects: LCSH: Cable-stayed bridges--Reliability. | Suspension bridges--Reliability.
Classification: LCC TG405 .R45 2021 (print) | LCC TG405 (ebook) | DDC 624.2/38--dc23
LC record available at https://lccn.loc.gov/2020052976
LC ebook record available at https://lccn.loc.gov/2020052977

ISBN: 978-0-367-77026-6 (hbk)
ISBN: 978-0-367-77276-5 (pbk)
ISBN: 978-1-003-17059-4 (ebk)

Typeset in Times
by SPi Global, India

Contents

Preface

In recent decades, bridge design and construction technologies have experienced a remarkable development, and numerous long-span bridges have been constructed or are currently under construction all over the world. A long-span bridge is the key junction of highways or railways and thus require a higher safety level. However, long-span bridges suffer from harsh environmental effects and complex loading conditions, such as heavier traffic loading, more significant wind load and more severe corrosion effects, as well as other natural disasters. Consequently these effects may result in changes in the structural mechanical behavior, which may change the dynamic characteristics and resistance of the bridge during its lifetime.

Cable-supported bridges, including cable-stayed bridges and suspension bridges, are the main types of long-span bridges. A cable-supported bridge has one or more pylons, from which cables support the bridge deck. This type of bridge is widely used in highways crossing gorges, rivers, and gulfs, due to their superior structural mechanical properties and aesthetic appearance. With the increase in the development of cable-supported bridges, on the other hand, the service performance of these types of bridges is facing great challenges. The cables connecting bridge decks and pylons provide high degrees of redundancy for the bridge structure, thereby making the structural system stronger and more robust. However, cables are particularly vulnerable to fatigue damage and atmospheric corrosion during the service period, which contributes to the risk of bridge failure.

The uncertainties in structural parameters also contribute to the system safety risk. Therefore, the influence of cable degradation and associated uncertainties on the system safety of cable-stayed bridges deserves investigation. System reliability evaluation of cable-supported bridges can provide a theoretical basis for the scheme of cable replacement, that is, when and which of the cables should be replaced by new cables. Due to human error and structural uncertainties, the cables of cable-supported bridges also have higher risk of failure during construction, which may lead to the damage or even collapse of the entire bridge. Since engineering structures involve uncertainties associated with the structural systems' initial imperfections, resistance deterioration, and increasing loads, probabilistic-based structural safety has received increasing attention in recent investigations.

Structural reliability is a probabilistic method that takes uncertain information in structural response analysis into consideration. Since environmental loads are mostly random in nature, the reliability theory is a powerful tool to assess the lifetime safety of these bridges. Cable-supported bridges have unique characteristics compared to short- to medium-span bridges, which makes the traditional reliability evaluation approaches of short-span bridges inappropriate for cable-supported bridges. The most critical factors are: first, the structural preference functions are mostly explicit; second, the structural mechanical behavior has higher nonlinearity; and finally, the bridge structure is statically indeterminate. Since engineering structures are mostly statically indeterminate structures consisting of various types of components, the failure modes exhibit randomness under random loads. A single failure of one component may not lead to the failure of the entire structure. For example, a cable-stayed bridge is composed of stay cables, girders, and towers. The potential failure modes are bending failure of girders, strength failure of cables, and displacement failure of girders. The failure of a cable may not lead to failure of the entire bridge. Thus, system reliability of these engineering structures is desirable and an important research topic. In addition, assessments of the fatigue, dynamic, and seismic reliability of engineering structures are important. These unique factors for cable-supported bridges make the reliability evaluation more complicated.

This book presents a comprehensive illustration for reliability and safety evaluation of cable-supported bridges. The book is composed of 11 chapters, each of which presents a complete investigation and case study with results. Advanced intelligent algorithms including neural networks and learning machines have been presented for both component- and system-level reliability evaluation of cable-supported bridges, aimed at accounting for structural nonlinearity more efficiently and accurately. In addition, the influence of cable corrosion and cable rupture on the system reliability of cable-stayed bridges is investigated and presented. Subsequently, fatigue reliability evaluation and fatigue crack propagation of steel bridge decks have been conducted in these reported studies, considering site-specific monitoring data for traffic and structural health. Finally, the maximum probability and dynamic traffic load effects on cable-supported bridges are evaluated. This book also intends to provide engineers and researchers an intuitive appreciation for probability theory, statistical methods, and reliability analysis methods. It is a good educational resource for engineering students at both undergraduate and graduate levels, as well as for practicing engineers.

Naiwei Lu
Associate Professor of Civil Engineering
Changsha University of Science and Technology

Yang Liu
School of Civil Engineering
Hunan University of Technology

Mohammad Noori
Department of Mechanical Engineering
California Polytechnic State University
San Luis Obispo

Acknowledgments

The majority of the work reported in this book formed part of various research projects conducted by the first two authors, which were funded by the National Science Foundation of China (Grant number 51908068), the National Science Foundation of Hunan Province (Grant numbers 2020JJ5140 and 2020JJ5589), the Innovation Platform Open Fund Project of Hunan Education Department (19K002), and the Industry Key Laboratory of Traffic Infrastructure Security Risk Management at Changsha University of Science and Technology (Grant numbers 19KF03 and 19KB02).

Under the supervision of the third author the first author also undertook postdoctoral research work for two years on several related topics that resulted in several joint papers and a two-volume book on topics related to the reliability and serviceability of long-span steel bridges.

The first author would like to express his heartfelt thanks to the group leader Prof. Jianren Zhang, and Prof. Yang Liu, Prof. Lei Wang, and Prof. Hui Peng, who provided the valuable opportunity to conduct the related research. The help from colleagues in the School of Civil Engineering at Changsha University of Science and Technology, and graduate students in the research group is also appreciated. Thanks to the anonymous professorial reviewers for their critical comments and careful guidance on improving the writing quality of the published articles. The first author would also like to thank his parents for their selfless love and care given from childhood. Finally, the first author would like to thank his lovely and virtuous wife Dr. Yuan Luo for her continued support and being a dedicated life companion. This book is a gift for the second birthday of the first author's cute baby girl.

The third author would like to express his sincere gratitude to his lifelong companion and wife, Nahid, without whose unconditional support, sacrifice and encouragements throughout their many years of companionship, participation in the writing of this book and his other scholarly work would not have been possible. He is also grateful for the opportunity to collaborate with Dr. Naiwei Lu, beginning around a decade ago when Dr. Lu worked with him as a postdoctoral researcher, and has since evolved into a long lasting scholarly partnership, a sincere collegial relationship, and friendship.

Notes on the Editors

Naiwei Lu is an associate professor of civil engineering at the Changsha University of Science and Technology, China. He received his B.D. (2008), M.D. (2011), and Ph.D. (2015) from Changsha University of Science and Technology. He was a postdoctoral researcher (May. 2015 to Jun. 2017) at the Southeast University, China, and a visiting postdoctoral researcher (Jan. 2016 to Jan. 2017) at the Institute for Risk and Reliability at Leibniz University, Hanover, Germany. His research interests are in reliability and safety assessment of long-span bridges. Email: lunaiweide@163.com

Yang Liu is a professor of civil engineering and an eminent scholar at Hunan University of Technology, and a guest professor of civil engineering at Changsha University of Science and Technology. He received his Ph.D. (2005) from Hunan University, China. His research interests are in bridge safety control and reliability assessment. Email: liuyangbridge2@163.com

Mohammad Noori is a professor of mechanical engineering at California Polytechnic State University, San Luis Obispo. He received his B.S. (1977), his M.S. (1980) and his Ph.D. (1984) from the University of Illinois, Oklahoma State University, and the University of Virginia respectively; all degrees in civil engineering with a focus on applied mechanics. His research interests are in stochastic mechanics, nonlinear random vibrations, earthquake engineering, and structural health monitoring, AI-based techniques for damage detection, stochastic mechanics, and seismic isolation. He serves as either executive editor, associate editor, technical editor, or member on the editorial boards of eight international journals. He has published over 250 refereed papers, has been an invited guest editor of over 20 technical books, has authored/co-authored six books, and presented over 100 keynote and invited presentations. He is a fellow of ASME, and has received the Japan Society for Promotion of Science Fellowship. Email: mnoori@outlook.com

List of Contributors

Michael Beer
Leibniz Universität Hanover

Yuan Luo
Hunan University of Technology

Haiping Zhang
Hunan University of Technology

Yafei Ma
Changsha University of Science and
Technology

Da Wang
Changsha University of Science and
Technology

Xinfeng Yin
Changsha University of Science and
Technology

Yang Deng
Beijing University of Civil Engineering
and Architecture

Xinhui Xiao
Hunan University of Technology

Qinyong Wang
Changsha University of Science and
Technology

Fanghui Chen
Changsha University of Science and
Technology

Bowen Wang
Changsha University of Science and
Technology

Kai Wang
Changsha University of Science and
Technology

Honghao Wang
Changsha University of Science and
Technology

Mengnan Yu
Changsha University of Science and
Technology

Jian Cui
Changsha University of Science and
Technology

Zhou Huang
Changsha University of Science and
Technology

Ming Zhang
Changsha University of Science and
Technology

Dan Zeng
Changsha University of Science and
Technology

List of Contributors

Minhua Bear
University of Science and Technology

Wei Liu
Hunan University of Technology

Xiaqing Xhang
Hunan University of Technology

Anhui Wu
Hunan University of Science and
Technology

Bo Yuan
Hunan University of Science and
Technology

Xiaomin Lu
Hunan University of Science and
Technology

Xing Deng
Hunan University of Science and
Machine Learning

Leshan Tan
Hunan University of Technology

Qingsong Wang
Hunan University of Science and
Technology

Fangjun Chen
Changsha University of Science and
Technology

Haoran Wang
Hunan University of Science and
Technology

Kai Yuan
Changsha University of Science and
Technology

Jianzhong Wang
Changsha University of Science and
Technology

Wenxiao Yu
Changsha University of Science and
Technology

Jian Sun
Changsha University of Science and
Technology

Zihui Huang
Changsha University of Science and
Technology

Xinru Zhang
Changsha University of Science and
Technology

Dan Zeng
Changsha University of Science and
Technology

List of Abbreviations

AASHTO	American Association of State Highway and Transportation Officials
ACI	American Concrete Institute
ADR	Assessment dynamic ratio
ADTT	Average daily traffic
AI	Artificial intelligence
AIC	Akaike information criterion
ANN	Artificial Neural Network
APDL	Algorithmic processor description language
ASVR	Adaptive support vector regression
B&B	Branch-and-bound
BP	Back propagation
CA	Cellular automaton
CDF	Cumulative distribution function
COV	Coefficient of variation
CSRA	Complex structural reliability analysis
DAF	Dynamic amplification factor
DBN	Deep belief network
DLA	Dynamic load allowance
DOF	Degrees of freedom
DPS	Data processing system
EDWL	Equivalent dynamic wheel load
EM	Expectation maximization
FE	Finite element
FEA	Finite element analysis
FEM	Finite element method
FORM	First Order Reliability method
FOSM	First-order-second-moment
GA	Genetic algorithm
GEV	Generalized extreme value
GMM	Gaussian mixture model
GVW	Gross vehicle weight
HLRF	Hasofer Lin and Rackwtz Fiesser
IIW	International Institute of Welding
IS	Importance sampling
LEFM	Linear elastic fracture mechanics
LIBSVM	Library for Support Vector Machines
LSF	Limit state function
LSSVR	Least squares support vector regression
MATLAB	Matrix Laboratory, which is a programming language
MCIS	Monte Carlo Importance Sampling
MCS	Monte Carlo simulation
MIDAS	Name of a finite-element program

MMP	Most probable point
MPFP	Most probable failure point
MSE	Mean square error
MSR	Matrix-based system reliability
MTS	Maximum tensile stress
OSD	Orthotropic steel decks
PDF	Probability density functions
PDF	Probability density functions
PNET	Probability network estimating technique
PSO	Particle swarm optimization
RADTT	Growth ratio of average daily truck volume
RBF	Radial basis function
RBM	Restricted Boltzmann machines
RFEA	Random finite element analysis
RGVW	Growth ratio of Growth vehicle weight.
RMS	Root-mean-square
RMSE	Root-mean-square error
RRC	Road roughness condition
RSM	Response surface method
SHM	Structural health monitoring
SIF	Stress intensity factor
SRM	Structural risk minimization
SVC	Support vector machine for classification
SVM	Support vector machine
SVR	Support vector regression
TBI	Traffic-bridge interaction
TC	Time period
TL	Lifetime of a bridge
UD	Uniform design
VBI	Vehicle-bridge interaction
WIM	Weigh-in-motion
WT	Wavelet Transfer

1 Introduction

Naiwei Lu

Changsha University of Science and Technology, China

Yang Liu

Hunan University of Technology, China

Mohammad Noori

California Polytechnic State University, USA

CONTENTS

1.1 SAFETY OF CABLE-SUPPORTED BRIDGES

With the steady expansion of the global economy, the transportation market is growing rapidly, especially in developing countries. The requirements for highways and railways are still critical, especially for mountainous and cross-sea areas. Therefore, numerous long-span bridges are currently under construction or have been built over the past decade. These major infrastructure expansions and projects directly drive development in the construction and design technologies. A long-span bridge is a key junction of highways or railways that requires a higher safety level. On the other hand, long-span bridges suffer from harsh environmental effects and complex loading conditions, such as heavier traffic loading, more significant wind load effects, more severe corrosion, and other natural disasters. These effects can result in changes in the structural mechanical behavior, which may change the dynamic characteristics and resistance of the bridge during its lifetime. Considering their rigorous service environments the lifetime safety evolution of these long-span bridges is an essential task.

Cable-supported bridges, including both cable-stayed and suspension bridges, are the main types of long-span bridges. A cable-supported bridge has one or more pylons, from which cables support the bridge deck. Cable-supported bridges are widely used in highways crossing gorges, rivers, and gulfs, due to their superior structural mechanical property and beautiful appearance. Some of the most aesthetically designed and famous cable-stayed bridges and suspension bridges are shown in Figure 1.1 and Figure 1.2, respectively. The information about the world famous largest cable-stayed bridges and suspension bridges is listed in Table 1.1 and Table 1.2, respectively.

With such rapid development of cable-supported bridges, on the other hand, their service performance is facing a great challenge. The cables in a cable-supported bridge connecting bridge decks and pylons are critical components ensuring the long-span capability of bridge girders. The cables provide high degrees of redundancy for the structure, thereby making the structural system stronger and more robust. However, cables are particularly vulnerable to fatigue damage and atmospheric corrosion during the service period, which contributes to the risk of bridge failure. A corrosive environment, fatigue damage, and vibration, affect the cables in cable-supported bridges where corrosion and degradation can be clearly seen in Figure 1.3. Strength degradation of the cables may lead to the collapse of the entire bridge. Stewart and Al-Harthy (2008) indicated that a spatial variability in corrosion of a steel bar led to a 200% higher failure probability. In practice, fatigue damage and corrosion are normal phenomena for steel strands or parallel wires in a stay cable. An inspection conducted by Mehrabi et al. (2010) showed that 39 of the 72 cables of the Hale Boggs Bridge were critically in need of repair or replacement after a 25-year service period. Although a cable-stayed bridge is designed with enough conservatism, the system is still vulnerable due to cable degradation.

1.2   UNCERTAINTIES IN CABLE-SUPPORTED BRIDGES

Engineering structures involve uncertainties associated with initial imperfections in the structural system, structural manufacturing errors, material characteristics, structural resistance deterioration, and external loads, such as traffic loads, wind loads, temperature effects, foundation settlements. Compared to short-to-medium span bridges, cable-supported bridges have unique characteristics, which makes the uncertainty quantification of cable-supported bridges more complicated. The most critical factors are complex performance functions, higher order nonlinearity, and statical indeterminacy. These unique factors for cable-supported bridges make the reliability evaluation more complicated.

Taking into account structural uncertainties and cable degradation, the system safety evaluation of cable-stayed bridges is more complicated. Each cable provides a degree of redundancy for a cable-stayed bridge; bridge failure can be defined as several components connected in series or in parallel. Probabilistic-based structural safety has received increasing attention in recent investigations. Uncertainties also exist in the external loads which are stochastic, such as traffic loads, wind loads, and temperature effects. The stochastic characteristics of these loads mostly depend on site-specific characteristics, which should be measured on site.

(a)

(b)

FIGURE 1.1 Long-span cable-stayed bridges: (a) Russky Bridge; (b) Sutong Yangtze River Bridge (Wiki).

(a)

(b)

FIGURE 1.2 Long-span suspension bridges: (a) Akashi Kaityo Bridge; (b) Xihoumen Bridge (Wiki).

TABLE 1.1
List of World-Famous Largest Cable-Stayed Bridges

No.	Bridge name	Span length (m)	Completed time	Country
1	Russky Bridge	1104	2012	Russia
2	Sutong Yangtze River Bridge	1088	2008	China
3	Stonecutter Bridge	1018	2009	China
4	Edong Yangtze River Bridge	926	2010	China
5	Tatara Bridge	890	1999	Japan
6	Pont de Normandie	856	1995	France
7	Jiujiang Yangtze River Bridge	818	2013	China
8	Jingyue Yangtze River Bridge	816	2010	China
9	Second Wuhu Yangtze River Bridge	806	2017	China
10	Incheon Bridge	800	2016	South Korea

TABLE 1.2
List of World-Famous Largest Suspension Bridges

Rank	Bridge name	Span length (m)	Completed time	Country
1	Akashi Kaityo Bridge	1991	1998	Japan
2	Xihoumen Bridge	1650	2009	China
3	Great Belt Bridge	1624	1998	Denmark
4	Osman Gazi Bridge	1550	2016	Turkey
5	Yi Sun-sin bridge	1545	2012	South Korea
6	Runyang Bridge	1490	2005	China
7	Dongting Lake Bridge	1480	2018	China
8	Nanjing Fourth Yangtze Bridge	1418	2012	China
9	Humber Bridge	1410	1981	U.K.
10	Yavuz Sultan Selim Bridge	1408	2016	Turkey

Structural reliability evaluation is a probabilistic method that considers uncertain information in structural response analysis. Since the environmental loads are mostly random in nature, reliability theory is a powerful tool to assess the lifetime safety of these bridges. In particular, fatigue, dynamic, and seismic reliability assessments of engineering structures have been widely studied in recent years (Lu and Noori 2018). Since engineering structures are mostly statically indeterminate structures consisting of various types of components, the failure modes exhibit randomness under random loads. The single failure of a component may not lead to failure of the entire structure. For example, a cable-stayed bridge is composed of stay cables, girders, and towers. The potential failure modes are bending failure of girders, strength failure of cables, and displacement failure of girders. The failure of a cable may not lead to failure of the entire bridge. Thus, applications of structural reliability theory to cable-stayed bridge structures are desirable and important research topics.

FIGURE 1.3   Corrosion of cables in cable-supported bridges.

1.3   SYSTEM RELIABILITY OF LONG-SPAN BRIDGES

Long-span bridges are statically indeterminate structures comprised of various types of components. Their failure modes are random and various, such as the bending failure of girder and pylons, cable rupture, fatigue damage, and first-passage failure.. Collapse of a cable-stayed bridge is caused by a sequence of critical components in series or in parallel. Thus, the structural system reliability theory is essential to assess the reliability and safety of long-span bridges.

For the system reliability assessment of these structures, two key issues need special consideration. First, the structural performance functions are implicit with high order nonlinear behaviors. The following three commonly used methods, Artificial Neural Network (ANN), Respond Surface Method (RSM) and Monte Carlo Simulation (MCS) have been developed to address these problems. Rocco (2002) introduced the SVM method into reliability analysis and proposed the SVM-MC method. Hurtado and Alvarez (2003) adopted SVM in conjunction with stochastic finite elements to analyze structural reliability. Gomes and Awruch (2004) demonstrated that ANN combined with MCS or First Order Reliability method (FORM) was an efficient algorithm to solve reliability problems with implicit response functions. Since the principle in ANN is based on the least mean square error, local convergence and over fitting are the key problems to be overcome when used to approximate the implicit functions. In order to address these problems, a support vector regression (SVR) with the principle of structural risk minimization (SRM) was proposed as an alternative approach. Guo and Bai (2009) introduced least squares support vector regression (LSSVR) in structural reliability analysis. However, the scope of the application and utilization of the SVM method in structural system reliability assessment has been relatively deficient.

Another main challenge to system reliability analysis is that there are numerous possible failure sequences. Without an efficient search scheme, an excessive number of failure sequences must be proposed to accurately estimate the failures probability. Furthermore, quantifying the likelihood of system-level failure sequences requires drastically new structural analyses in order to account for load redistributions and various uncertainties. The MCS method is the most straightforward approach among the existing system reliability analysis methods. Several non-sampling-methods have been developed to increase the computational efficiency of the transitional MCS method. The widely used non-sampling-method selectively searches schemes based on an event-tree of potential failure sequences, such as the branch-and-bound (B&B) method developed by Murotsu (1984). In order to overcome the time-consuming problem and the possible misidentification of critical failure sequences in the B&B method, Lee and Song (2011) developed a new B&B method for FE-based system reliability analysis of continuum structures by modifying the LSFs. Kang et al. (2008) proposed a matrix-based system reliability (MSR) method that generalizes the linear programming bounds method to represent the system events in the case of complete information. Although extensive efforts have been made to develop structural system reliability assessments, the trade-off between time-consuming computation and efficiency is still a challenging issue.

1.4 TIME-VARYING RELIABILITY OF BRIDGES DURING CONSTRUCTION

Due to human error and structural uncertainties, the cables of cable-stayed bridges have a higher risk of failure during construction which may lead to the damage or even collapse of the bridge. Structural uncertainties during construction add to the complication of the reliability evaluation of cable-supported bridges. In addition, the structural mechanical behavior during is different from that of the in-service stage. Most long-span bridges are statically indeterminate, while their construction states are determinate.

Numerous efforts have been made to research the uncertainties in bridge construction stages. Cheng and Li (2009) took the static wind load-related parameters as variables and evaluated the reliability of a steel arch bridge under wind load. Cho and Kim (2008) evaluated the risks in a suspension bridge by considering an ultimate limit state for the fracture of main cable wires during construction phases. Cheng and Xiao (2005) evaluated the serviceability reliability of a cable-stayed bridge by utilizing a combination of the response surface, finite element, and first order reliability methods. However, little attention has been paid to deal with the uncertainties and reliabilities of cable-stayed bridges during construction, where the cables and girders are critically important during the cantilever state. The vulnerability of cable-stayed bridges subject to cable loss scenarios during construction has neither been thoroughly investigated nor recommended in the current studies and design codes. The structural safety of cable-stayed bridge caused by cable failure is worthy of research.

1.5 FATIGUE RELIABILITY OF STEEL BRIDGES

Long-span steel bridges are vulnerable to repeated loads caused by traffic, wind, gusts, and the changing environment. These combined effects can lead to complex modes of fatigue failure. Fatigue damage is one of the main forms of deterioration for structures and can be a typical failure mode due to the accumulation of damage.

Steel decks are widely used in long-span bridges due to the advantages of light weight, high strength, and short construction period (Wang et al. 2010; Chen et al. 2011a; Fu et al. 2018). However, recent field investigations on several collapsed steel bridges (Han et al. 2015; Lu et al. 2017) indicated that fatigue damage induced by accidentally overloaded trucks contributed to the bridge failures. Thus increasing traffic loads may become a safety hazard for the fatigue safety of steel bridges especially in developing countries (Guo and Chen 2013). Therefore, integrating actual traffic information into the fatigue reliability evaluation of existing steel bridges is of great importance as it will provide a more accurate evaluation result and a theoretical basis for transportation management. Both deterministic and probabilistic procedures have been applied to estimate the fatigue damage of structures. Since the major cause of fatigue in steel bridges is vehicle load, which is a strong stochastic process, fatigue reliability evaluation has resulted in more research studies into this aspect. Chen et al. (2011b) assessed fatigue reliability of Tingma Bridge under multi-loadings based on an SHM system. Zhang et al. (2012) presented a comprehensive framework for fatigue reliability estimation of bridges under combined dynamic loads from vehicles and wind. Guo et al. (2012a) proposed an advanced traffic load model factoring in the uncertainties associated with the number of axles, and axle spacing and weights. Kwon and Frangopol (2011) integrated the fatigue reliability and crack growth models, with the probabilistic detection data for fatigue assessment and management of existing bridges.

A critical step for measuring fatigue damage of steel bridges is the simulation of both structural welded details and fatigue truck load models (Dung et al. 2015). The configuration of trucks greatly contributes to the fatigue stress histories of bridge decks (Deng et al. 2011). Thus, numerous research efforts have been reported on the fatigue truck load model. Laman and Nowak (1996) developed two fatigue truck load models including a tree-axle truck and a four-axle truck based on the weigh-in-motion (WIM) data of five steel bridges, which were verified and compared with measured results. Subsequently, several fatigue truck load models have been presented in native design codes, such as the American Association of State Highway and Transportation Officials. Recently, a multi-loading model, with consideration of railway, highway, and wind loads, for the fatigue reliability assessment of Tingma Bridge was developed by Chen et al. (2011b). However, Chen's vehicle model simplified the vehicles as a concentrated force, where the multi-axle effect and the dynamic response are ignored. In response to this limitation, Zhang et al. (2013, 2014) presented a comprehensive framework for fatigue reliability estimation of bridges subjected to combined dynamic loads from vehicles and wind loads, where the vehicle model is more realistic and the vehicle-bridge interactions were taken into account. The influence of dynamic impacts due to deteriorated road surface conditions and vehicle suspension systems was investigated by MacDougall et al. (2006) and Zhang and Cai (2011). In order to conduct a probabilistic analysis, Guo et al. (2012b) developed an advanced fatigue vehicle model accounting for the uncertainties associated with the number of axles, axle spacing, and axle weights by using WIM data. Given the need for a more realistic model for the stochastic traffic model, which incorporates an appropriate PDF of axle weight, Xia et al. (2012) implied mixture distribution models with an expectation maximization (EM) algorithm and the Akaike information criterion (ACI) to establish PDFs of fatigue stress ranges. Even though fatigue truck load models have been investigated and adopted by several fatigue design specifications, limited efforts have been made into research on probabilistic modeling of a general fatigue truck loads model, as this apparently has a significant influence on the fatigue reliability of steel bridges. Therefore, more research needs to be undertaken in this area..

1.6   DYNAMIC RELIABILITY OF BRIDGES UNDER VEHICLE LOADS

Compared with short-span bridges, long-span bridges exhibit unique features such as higher traffic volume, simultaneous presence of multiple vehicles, and sensitivity to strong wind excitations. Wind loads and vehicle loads are two main continuous variable loads for long-span in-service bridges. As stated earlier, stochastic traffic flow should be considered in the analysis of long-span bridges since random loading caused by traffic flow results in a direct and severe vibration of bridges, compared with, and in contrast to, the transient vibration of a single vehicle. Regarding the probabilistic model of vehicles, WIM technology has been used in the statistical analysis of vehicle characteristics such as vehicle speed, axle weight, and vehicle type (Schadschneider 2002). Often, the MCS approach has been adopted to simulate a similar stochastic traffic flow with consideration of probability of vehicle type, axle weight, vehicle distance, and vehicle speed. A comprehensive MCS method for free-flowing traffic was presented and demonstrated by measuring the data of five European highway bridges (Enright and O'Brien 2013). In order to model the vehicle state more realistically, a cellular automation-based traffic flow simulation technique was proposed to simulate the stochastic live load from traffic for long-span bridges (Chen and Wu 2011).

Research work on the interaction between vehicles and bridges originated in the middle of the 20th century. To start off with, the vehicle loads were modeled as a constantly moving force, moving mass, or moving mass-spring. Further progress in this research area led to a fully computerized approach for assembling equations of the motions of coupled vehicle-bridges, which was proposed by modeling the vehicles as a combination of a number of rigid bodies connected by a series of springs and dampers (Ahmari et al. 2015). On this basis, a 3D simulation approach including a 3D suspension vehicle model and a 3D dynamic bridge model was developed. The current AASHTO specifications (AASHTO 2007) define the dynamic effects duo to moving vehicles by impact factors attributed to hammering effect and road roughness. The road roughness could be assumed as a zero-mean stationary Gaussian random process and could be generated through an inverse Fourier transformation. It was later shown that the foundation settlement and other environmental factors would affect the bridge-vehicle interaction due to the shape of deck (Chen and Wu 2010).

Without considering wind dynamic impacts on the vehicles, the dynamic wheel load would be underestimated by about 6–11% (Cai and Chen 2004). When considering the wind excitations, the vehicle-bridge interaction is more prominent and complex, and a large body of research has been conducted on vehicle-wind-bridge interaction. A comprehensive framework regarding a vehicle-wind-bridge dynamic analysis of coupled 3D was first presented by Chen and Cai (2007). In their framework, a series of vehicles consisting of different numbers and types of vehicles driving on bridges under hurricane-induced strong winds was included. Based on that framework an equivalent dynamic wheel load (EDWL) approach and the CA traffic simulation were adopted to analyze the dynamic performance of long-span bridges under combined loads of stochastic traffic and wind excitations (Wu and Chen 2011). A reasonable framework to replicate probabilistic traffic flow, characterize the dynamic interaction, and assess the structural performance under strong wind and heavy traffic was presented to study the probabilistic dynamic behavior of long-span bridges under extreme events (Rice 1945).

Applications of first-passage reliability to engineering structures are very interesting since safety assessment and design can be proposed to guarantee structural safety (Cai and Lin 1994). Significant progress in structural reliability evaluation has been achieved in the last decades utilizing nonlinear stochastic structural dynamics. The dynamic reliability approach should be utilized to estimate reliability since this analysis incorporates random vibration theory and the effect of bridge-vehicle interactions caused by random traffic flow is considered.

Few research works have been reported in the literature on the probabilistic dynamic analysis of long-span bridges subjected to combined stochastic traffic and wind excitations, which is extremely important for the safety of long-span bridges. Furthermore, due to the nature of time-varying differential equations in the interaction system, most dynamic analyses reported in this area have focused on time domain analysis, while very limited development has occurred in the frequency domain. However, incorporating random vibration in the aforementioned coupled system, which requires and necessitates the use of spectral analysis, is more important and results in more valuable information in the frequency domain. In bridge engineering, the first-passage reliability method is suitable for the safety assessment of existing bridges under vehicle loads.

1.7 PROBABILISTIC TRAFFIC LOAD EFFECTS ON CABLE-SUPPORTED BRIDGES

The live load specified in the national design specification is conventionally evaluated based on site-specific traffic measurements collected several decades ago. For instance, the live load in the Load and Resistance Factor Design code of America was evaluated based on data collected in Ontario in the 1970s (Nowak 1994). The live load in Eurocode 3 was made based on weeks of data collected in Auxerre in the 1980s (ECS 2005). China's code (MOCAT 2015) was made based on five highway traffic measurements in the 1990s. However, Fu and You (2009) indicated that four out of 1319 trucks in China cause larger bending moments of short-span bridges than that evaluated according to the national design specification. Long-span bridges suffer from the simultaneous presence of multiple vehicle loads compared to short-to-medium span bridges. Therefore, evaluation of maximum traffic load effects on long-span bridges using actual traffic data deserves investigation.

In addition to the random nature of the environmental loads, it is important to consider time-varying properties, such as traffic growth in volume and gross weight, fatigue damage accumulation, and deterioration of cable strength. For instance, the current traffic load has increased significantly in recent years, which may prove hazardous for serviceability, fatigue, or even the safety of existing bridges. In fact, several short- and medium-span bridges have collapsed due to extremely overloaded trucks. The growing traffic load impacts the fatigue damage accumulation on the bridge deck and the maximum traffic load effect. Furthermore, a single failure of component may not lead to failure of the entire structure. For example, a cable-stayed bridge is composed of stay cables, girders, and towers. The potential failure modes are bending failure of girders, strength failure of cables, and displacement failure of girders. Finally, cables are particularly vulnerable to fatigue damage and atmospheric corrosion during the service period, which then contributes to the risk of bridge failure.

These important changes that need to be better understood are time variant phenomena. The safety performance of these long-span suspension bridges is facing numerous threats, which are mostly random in nature. All these loading conditions need to be considered and must satisfy design criteria.

1.8 CONTENTS OF THIS BOOK

This book presents a comprehensive illustration for the reliability and safety evaluation of cable-supported bridges. The book is composed of ten chapters each of which presents a complete investigation achievement. Advanced intelligent algorithms, including neural networks and learning machines, are presented for both component- and system-level reliability evaluation of cable-supported bridges, aimed at accounting for structural nonlinearity more efficiently and accurately. In addition, the influence of cable corrosion and rupture on the system reliability of cable-stayed bridges was investigated. Subsequently, fatigue reliability evaluation and fatigue crack propagation of steel bridge decks were conducted considering site-specific traffic monitoring data and structural health monitoring data. Finally, the maximum probability and dynamic traffic load effects on cable-supported bridges were evaluated. This book also intends to provide engineers and researchers an intuitive appreciation for probability theory, statistical methods, and reliability analysis methods. It is a suitable educational resource for engineering students at both undergraduate and graduate levels, as well as for practicing engineers.

REFERENCES

Ahmari, S., Yang, M. and Zhong, H. 2015. Dynamic interaction between vehicle and bridge deck subjected to support settlement. *Engineering Structures* 84: 172–183.

American Association of State Highway and Transportation Officials (AASHTO). 2007. *LRFD bridge design specifications*, Washington, DC.

Cai, C. S. and Chen, S. R. 2004. Framework of vehicle–bridge–wind dynamic analysis. *Journal of Wind Engineering and Industrial Aerodynamics* 92(7): 579–607.

Cai, G. Q. and Lin, Y. K. 1994. On statistics of first-passage failure. *Journal of Applied Mechanics* 61(1): 93–99.

Chen, S. R. and Cai, C. S. 2007. Equivalent wheel load approach for slender cable-stayed bridge fatigue assessment under traffic and wind: Feasibility study. *Journal of Bridge Engineering* 12(6): 755–764.

Chen, S. R. and Wu, J. 2010. Dynamic performance simulation of long-span bridge under combined loads of stochastic traffic and wind. *Journal of Bridge Engineering* 15(3): 219–230.

Chen, S. R. and Wu, J. 2011. Modeling stochastic live load for long-span bridge based on microscopic traffic flow simulation. *Computers and Structures* 89(9): 813–824.

Chen, Z. W., Xu, Y. L. and Wang, X. M. 2011a. SHMS-based fatigue reliability analysis of multiloading suspension bridges. *Journal of Structural Engineering* 138(3): 299–307.

Chen, Z. W., Xu, Y. L., Xia, Y., Li, Q. and Wong, K. Y. 2011b. Fatigue analysis of long-span suspension bridges under multiple loading: Case study. *Engineering Structures* 33(12): 3246–3256.

Cheng, J. and Li, Q. 2009. Reliability analysis of a long span steel arch bridge against wind-induced stability failure during construction. *Journal of Constructional Steel Research* 65(3): 552–558.

Cheng, J. and Xiao, R. C. 2005. Serviceability reliability analysis of cable-stayed bridges. *Structural Engineering and Mechanics* 20(6): 609–630.

Cho, T. and Kim, T. S. 2008. Probabilistic risk assessment for the construction phases of a bridge construction based on finite element analysis. *Finite Elements in Analysis and Design* 44(6): 383–400.

Deng, Y., Ding, Y. L. and Li, A. Q. et al. 2011. Fatigue reliability assessment for bridge welded details using long-term monitoring data. *Science China Technological Sciences* 54(12): 3371–3381.

Dung, C. V., Sasaki, E. Tajima, K. and Suzuki, T. 2015. Investigations on the effect of weld penetration on fatigue strength of rib-to-deck welded joints in orthotropic steel decks. *International Journal of Steel Structures* 15(2): 299–310.

Enright, B. and O'Brien, E. J. 2013. Monte Carlo simulation of extreme traffic loading on short and medium span bridges. *Structure and Infrastructure Engineering* 9(12): 1267–1282.

European Committee for Standardization (ECS). 2005. *Eurocode 3: Design of steel structures: general rules and rules for buildings*, EN 1993-1-1, Brussels, Belgium.

Fu, Z., Ji, B. Zhang, C. and Li, D. 2018. Experimental study on the fatigue performance of roof and U-rib welds of orthotropic steel bridge decks. *KSCE Journal of Civil Engineering* 22(1): 270–278.

Fu, G. and You, J. 2009. Truck loads and bridge capacity evaluation in China. *Journal of Bridge Engineering* 14(5): 327–335.

Gomes, H. M. and Awruch, A. M. 2004. Comparison of response surface and neural network with other methods for structural reliability analysis. *Structural safety* 26(1): 49–67.

Guo, Z. and Bai, G. 2009. Application of least squares support vector machine for regression to reliability analysis. *Chinese Journal of Aeronautics* 22(2): 160–166.

Guo, T. and Chen, Y. W. 2013. Fatigue reliability analysis of steel bridge details based on field-monitored data and linear elastic fracture mechanics. *Structure and Infrastructure Engineering* 9(5): 496–505.

Guo, T., Frangopol, D. M. and Chen, Y. 2012a. Fatigue reliability assessment of steel bridge details integrating weigh-in-motion data and probabilistic finite element analysis. *Computers and Structures* 112: 245–257.

Guo, T., Frangopol, D. M. and Chen, Y. 2012b. Fatigue reliability assessment of steel bridge details integrating weigh-in-motion data and probabilistic finite element analysis. *Computers and Structures* 112: 245–257.

https://en.wikipedia.org/wiki/List_of_longest_cable-stayed_bridge_spans.

https://en.wikipedia.org/wiki/List_of_longest_suspension_bridge_spans.

Han, Y., Liu, S., Cai, C. S., Zhang, J., Chen, S. and He, X. 2015. The influence of vehicles on the flutter stability of a long-span suspension bridge. *Wind & Structures: An International Journal* 20(2): 275–292.

Hurtado, J. E. and Alvarez, D. A. 2003. Classification approach for reliability analysis with stochastic finite-element modeling. *Journal of Structural Engineering* 129(8): 1141–1149.

Kang, W. H., Song, J. and Gardoni, P. 2008. Matrix-based system reliability method and applications to bridge networks. *Reliability Engineering & System Safety* 93(11): 1584–1593.

Kwon, K. and Frangopol, D. M. 2011. Bridge fatigue assessment and management using reliability-based crack growth and probability of detection models. *Probabilistic Engineering Mechanics* 26(3): 471–480.

Laman, J. and Nowak, A. 1996. Fatigue-load models for girder bridges. *Journal of Structural Engineering* 122(7), 726–733.

Lee, Y. J. and Song, J. 2012. Risk analysis of fatigue-induced sequential failures by branch-and-bound method employing system reliability bounds. *Journal of Engineering Mechanics* 137: 807–821.

Lu, N. and Noori, M. 2018. *Multi-scale reliability and serviceability assessment of in-service long-span bridges.* Momentum Press, New York.

Lu, N. W., Noori, M. and Liu, Y. 2017. Fatigue reliability assessment of welded steel bridge decks under stochastic truck loads via machine learning. *Journal of Bridge Engineering* 22(1): 04016105.

MacDougall, C., Green, M. and Shillinglaw, S. 2006. Fatigue damage of steel bridges due to dynamic vehicle loads. *Journal of Bridge Engineering* 11(3), 320–328.

Mehrabi, A. B., Ligozio, C. A., Ciolko, A. T. and Wyatt, S. T. 2010. Evaluation, rehabilitation planning, and stay-cable replacement design for the Hale Boggs Bridge in Luling, Louisiana. *Journal of Bridge Engineering* 15: 364–372.

Ministry of Communications and Transportation. (MOCAT) 2015. *General code for design of highway bridges and culverts GB 60–2015*, People's Communications Publishing Co.. Ltd, Beijing, China.

Murotsu, Y. 1984. Automatic generation of stochastically dominant modes of structural failure in frame. *Structural Safety* 2(1): 17–25.

Nowak, A. S. 1994. Load model for bridge design code. *Canadian Journal of Civil Engineering* 21(1): 36–49.

Rice, S. O. 1945. Mathematical analysis of random noise. *Bell System Technical Journal* 24(1): 46–156.

Rocco, C. M. 2002. Moreno JA. Fast Monte Carlo reliability evaluation using support vector machine. *Reliability Engineering and System Safety* 76(3): 237–243.

Schadschneider, A. 2002. Traffic flow: a statistical physics point of view. *Physica A: Statistical Mechanics and its Applications* 313(1): 153–187.

Stewart, M. G. and Al-Harthy, A. 2008. Pitting corrosion and structural reliability of corroding RC structures: Experimental data and probabilistic analysis. *Reliability Engineering & System Safety* 93: 373–382.

Wang, Y., Li, Z. X. and Li, A. Q. 2010. Combined use of SHMS and finite element strain data for assessing the fatigue reliability index of girder components in long-span cable-stayed bridge. *Theoretical and Applied Fracture Mechanics* 54(2): 127–136.

Wu, J. and Chen, S. R. 2011. Probabilistic dynamic behavior of a long-span bridge under extreme events. *Engineering Structures* 33(5): 1657–1665.

Xia, H., Ni, Y. Wong, K. and Ko, M. 2012. Reliability-based condition assessment of in-service bridges using mixture distribution models. *Computers and Structures* 106(9): 204–213.

Zhang, W. and Cai, C. S. 2011. Fatigue reliability assessment for existing bridges considering vehicle speed and road surface conditions *Journal of Bridge Engineering* 17(3): 443–453.

Zhang, W., Cai, C. S. and Pan, F. 2012. Fatigue reliability assessment for long-span bridges under combined dynamic loads from winds and vehicles. *Journal of Bridge Engineering* 18(8): 735–747.

Zhang, W., Cai, C. S. and Pan, F. 2013. Fatigue reliability assessment for long-span bridges under combined dynamic loads from winds and vehicles. *Journal of Bridge Engineering* 18(8): 735–747.

Zhang, W., Cai, C. S. Pan, F. and Zhang, Y. 2014. Fatigue life estimation of existing bridges under vehicle and non-stationary hurricane wind. *Journal of Wind Engineering and Industrial Aerodynamics* 133: 135–145.

2 Serviceability Reliability Assessment of Prestressed Concrete Cable-stayed Bridges Using Intelligent Neural Networks

Yang Liu
Hunan University of Technology, China

Naiwei Lu
Changsha University of Science and Technology, China

Qinyong Wang
Changsha University of Science and Technology, China

CONTENTS

2.1 INTRODUCTION: BACKGROUND

A cable-stayed bridge is composed of pylons, girders, and cables, which work as a system to support external loads. The cable-stayed bridge is widely used for long-span bridges because of its superior balance on long-span capability and relatively lower cost (Arangio and Bontempi 2015). In general, there are two types of girder for a cable-stayed bridge, namely, steel box girders and prestressed concrete girders. The latter are more suitable for the cable-stayed bridge with a maximum main span length of 400 m (Wang et al. 2019). However, several prestressed concrete cable-stayed bridges have collapsed or been severely damaged due to environment corrosion and truck overloading. Thus, the service reliability and safety of the cable-stayed bridge deserves investigation.

In general, the commonly used methods for structural reliability evaluation are the first-order second-moment (FOSM), Monte Carlo simulation (MCS), response surface (RSM) and random finite element analysis (RFEA) methods (Liu et al. 2016). Several advanced algorithms and computational software have been developed for the reliability assessment of long-span bridges (Chehade and Younes 2020). For instance, Chen et al. (2000) evaluated the reliability of the main girder for the Second Nanjing Bridge accounting for the geometric nonlinear effect of the cable-stayed bridge based on the sequence response surface method (RSM). Frangopol and Imai (2004) presented a system reliability approach implemented in a probabilistic finite element nonlinear elastic analysis applied for the Honshu–Shikoku bridge. Cheng and Xiao (2004) concluded that the sag effect of the stay cables cannot be neglected due to the geometric nonlinear effect. However, the beam-column effect and large displacement effect cannot be neglected for the static reliability evaluation of a cable-stayed bridge. Catbas et al. (2008) evaluated the reliability of a long-span truss bridge based on structural health monitoring data. Cheng et al. (2008) proposed a hybrid algorithm combining the RSM and neural networks (ANNs) to evaluate the static reliability of a cable-stayed bridge. Yan and Chang (2009) developed a stochastic finite element method to evaluate the cable-stayed bridge. Kong et al. (2012) investigated the influence of the vehicle load on the reliability index of a long-span bridge considering different mean coefficients of critical random variables. Deng et al. (2018) evaluated the fatigue reliability of the Runyang Yangtze River Bridge using the FOSM method. These advanced approaches were verified feasible and with reasonable accuracy and efficiency for the reliability assessment of long-span bridges.

For reliability evaluation of long-span cable-stayed bridges, there are three major problems needing special attention. First, the structural geometrical nonlinearity gradually increases with the increase of the bridge's main span length (An et al. 2016; Yang et al. 2017). Thus, structural nonlinearity in the limit state function should be accurately taken into consideration. Second, the cable-stayed bridge is a complex system consisting of cables, girders, and pylons which all add to the structural degree of indeterminacy (Cheng and Li 2009). Therefore, identifying the dominant failure mode is critical for a reliability analysis. Finally, the cables are vulnerable components leading to the degradation of the bridge system. Thus, the bridge parameters should be updated according to the bridge detection data, and time-varying reliability analysis is necessary.

This chapter presents a hybrid intelligent approach for the reliability assessment of long-span cable-stayed bridges based on artificial neural networks. The structural nonlinearity was considered in the kernel radial basis functions, and the system behavior was captured by the networks. The proposed hybrid approach was verified in two numerical studies and used for the case study of a long-span prestressed concrete cable-stayed bridge. Parametric studies of the bridge were conducted to reveal structurally sensitive parameters.

2.2 MATHEMATICAL MODELS

The structural system of cable-stayed bridges is highly statically indeterminate and geometrically nonlinear. Therefore, the limit state function of a long-span cable-stayed bridge is complex and implicitly associated with many random variables. Accordingly, it is important to establish an appropriate mathematical model to define the functional failure of the cable-stayed bridge.

In general, the safety factor of a bridge for the serviceability limit state is lower than that of the ultimate limit state (Wu and Zhao 2006). Assessment of the service performance of a bridge is critically analyzed under serviceability limit states to control stresses, cracks, and deformation. The role of the serviceability state is usually to guarantee the integrity of the bridge structure. Therefore, in a sense, the serviceability limit state of a bridge is critical to reflect a bridge's safety. In this study, the serviceability function is considered as the strength failure of a stay cable and the critical girder exceeding the deflection threshold of the main girder. The limit state functions of a cable-stayed bridge caused by the strength failure of cables and girder deflections are written as

$$Z_1 = T_u^i - T_{cab}^i \left(x_1, \ldots, x_n \right) \tag{2.1}$$

$$Z_2 = u_{max} - u_{mid} \left(x_1, \ldots, x_n \right) \tag{2.2}$$

where x are random variables, T_u^i is the yield strength of the ith stay cable, n is the number of strands in a stay cable, A is the cross-sectional area of single strands, σ_b is the yield strength of single strands, $T_{cab}^i (x_1,\ldots,x_n)$ is the axial force of the ith cable, u_{max} is the maximum vertical deflection for the concrete girders, which should not exceed a threshold of $L/500$ (L is the mid-span length of the main girder) under vehicle loading according to the design specification, $u_{mid}(x_1,\ldots,x_n)$ is the actual vertical displacement of the mid-span girders under actual vehicle loading. It is acknowledged that $T_{cab}^i(x_1,\ldots,x_n)$ and $u_{mid}(x_1,\ldots,x_n)$ are both high-order nonlinear implicit performance functions which can be expressed by the RSM, the Tylor expansion, and an artificial neural network. Accordingly, the innovation of this study is the utilization of a new intelligent neural network to approximate the implicit performance functions.

2.3 PROPOSED COMPUTATIONAL FRAMEWORK

In order to evaluate the reliability of cable-stayed bridges more accurately, a computational framework for a hybrid algorithm is presented to integrate the intelligent neural networks and the iterative solution steps. The hybrid algorithm is a combination of the finite element analysis (FEA), the neural network and the genetic

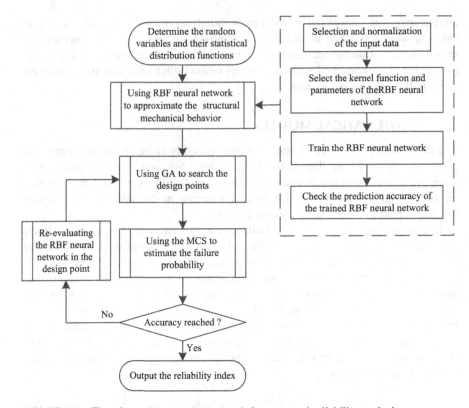

FIGURE 2.1 Flowchart of the hybrid approach for structural reliability analysis.

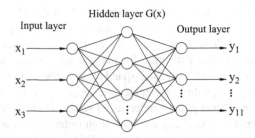

FIGURE 2.2 Structural diagram of the RBF neural network.

algorithm (GA), and the MCS. The computational flowcharts of the hybrid algorithm are shown in Figure 2.1. Detailed procedures are summarized below.

The RBF neural network in the hybrid algorithm was utilized to approximate the structural mechanical behavior. It is a feedforward neural network composed of an input layer, a hidden layer, and an output layer as shown in Figure 2.2, where x and y are the input data and output data of the structural response, respectively. The hidden layer of the radial basis function is the Gaussian function, whereas the output layer is a linear function. The kernel Gaussian is

$$G_i(x) = \exp\left[-\frac{(x-c_i)^T(x-c_i)}{2\sigma_i^2}\right] \qquad (2.3)$$

where x is an m-dimensional vector of input data, c_i and σ_i are the mean and standard deviation of the RBF neural network, respectively, and T is a transposed matrix.

For the finite element modeling of a long-span PC cable-stayed bridge, the high-order statically indeterminate and highly nonlinear implicit performance function should be carefully considered. The RBF neural network was used to approximate the structural response instead of the FEA model. The advantage of the RBF method is its computational efficiency and precision compared to traditional back propagation (BP) neural networks. The RBF neural networks are particularly efficient for structurally high-order multivariate nonlinear functions.

Sampling points were chosen according to the sample value range using a uniform design sampling method in order to ensure that the design sample points are within an effective range, which is recommended as μ-3σ and $\mu + 3\sigma$ according to the 3-sigma principle (Zhang et al. 2008). The uniform design method was used to simulate the experimental data because it has superior performance compared to the orthogonal design approach. A data processing system (DPS) (Tang and Zhang 2013) was used to simulate the uniform distribution samples. Finally, the RBF neural network was used to approximate the response function.

In order to search the highest probability point (or design point), the GA was utilized to optimize the searching scheme. Searching the design point can be simplified as a constrained optimization model. In general, the constrained optimization problem can be transformed into an unconstrained optimization function by introducing a GA function. Herein, the constraint optimization function can be written as

$$\begin{cases} \min \beta^2 = \sum_{i=1}^{n}\left[\frac{(x_i^* - \mu_{x_i}')}{\sigma_{x_i}'}\right]^2 \\ \text{s.t.} Z = Z_i(x_1,\ldots,x_n) = 0 \end{cases} \qquad (2.4)$$

where $x_1,\ldots, x_n$ are structurally independent random variables, x_i^* is the most probable failure point, $Z_i(x_1,\ldots,x_n)$ is the structural limit state function, μ_{xi}' and σ_{xi}' are the equivalent mean value and standard deviation, and β is the reliability index. For the non-Gaussian distribution variables, the Rosenblatt transformation or orthogonal transformation (Li and Hu 2000) can be used to transform the relative random variables into linear, independent, standard normal distribution random variables. The reliability index is defined as the shortest distance from the origin to the limit state plane in the standard normal coordinate system.

The failure probability P_f is usually extremely small for engineering structures, which leads to the computation difficulty. In general, if P_f is less than 10^{-3}, the sampling number should be greater than 10^5 to ensure computational accuracy. When the estimated error of the failure probability is less than 20%, the probability is greater than 95%, where the relative error ε of P_f is 0.2 and the significance level $u_{\alpha/2}$ is 1.96 (Gong et al. 2012). On this basis, the sampling number N should satisfy the following condition:

$$N \geq \frac{100\left[\Phi(-2\beta)\exp(\beta^2)-\Phi^2(-\beta)\right]}{\Phi^2(-\beta)-\Phi^3(-\beta)} \tag{2.5}$$

where β is a predicted reliability index and $\Phi()$ is the cumulative probability distribution function of a standard Gaussian function.

In order to reduce the sampling number, the importance sampling technology, also known as the variance reduction technique, can be adopted, as in certain conditions it improves the computational efficiency to guarantee the same accuracy. The importance sampling method increases the number of sample points in the failure region appropriately to reduce the variance by changing the sampling center. This method is often applied in many areas because of its feasibility and efficiency.

According to the aforementioned computational framework of the hybrid algorithm, a flowchart combining software including the DPS, neural networks, the GA toolbox, and the APDL language platform are summarized in the platform of MATLAB. The flowchart of the program is shown in Figure 2.3.

Initially, the DPS program is used to generate uniformly distributed samples that will be utilized as the input data for training neural networks. Subsequently, conduct the

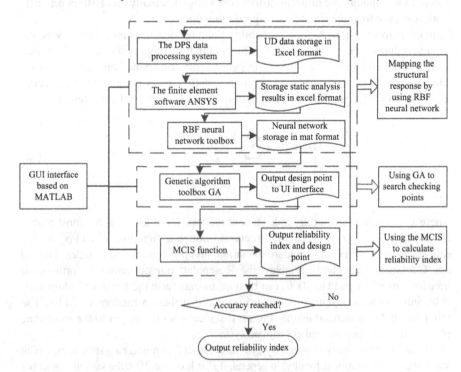

FIGURE 2.3 Software framework for structural reliability assessment.

FEA using commercial software to compute the bridge response as the output data corresponding to the input data. Third, the RBF neural network toolbox in the MATLAB can be utilized to train the networks. Note that the parameters in the toolbox should be appropriately selected to make the network more accurately. Finally, the structural reliability index β obtains its design point location by MCIS such that the precision requirement is satisfied. Otherwise, the above procedures are repeated until accuracy is reached.

2.4 VALIDATION EXAMPLES

In this section, the feasibility and computational accuracy will be verified by two numerical examples. The first example is an explicit complex performance function with a high degree of nonlinearity. The second example is a simplified structure of a cable-stayed bridge. In addition, the computational results were compared with other methods.

2.4.1 NUMERICAL EXAMPLE ANALYSIS

The first numerical example is an explicit performance function with two random variables. The performance function is $Z = 18.46-7.48X_1/X_2^3$, where X_1 and X_2 are both random variables with $X_1 \sim N(10,2)$ and $X_2 \sim N(2.5,0.375)$.

The second numerical example is an explicit performance function with three random variables. The performance function is $Z = X_1X_2-X_3$, where X_1 and X_2 are random variables with $X_1 \sim N(0.5472,0.0274)$, $X_2 \sim N(3.8,0.304)$, and $X_3 \sim N(1.3,0.91)$.

The numerical results were compared with the common approaches and the MCS results treated as the exact solution. The computational results are summarized in Table 2.1.

As observed in Table 2.1, all of the approaches can provide relative accurate results for the two numerical examples. However, the proposed hybrid algorithm provided more accurate reliability index with fewer iterations. Therefore, the hybrid algorithm can provide the same computational accuracy while reducing computational effect by approximately 20%.

2.4.2 THE BROTONNE CABLE-STAYED BRIDGE

The Brotonne cable-stayed bridge (Bruneau 1992) was selected as an additional verification example. The structural dimensions and failure modes are shown in Figure 2.3. More detailed information regarding the bridge can be found in Bruneau

TABLE 2.1
Computational Results of the Two Numerical Examples

Cases	Items	MCS	FOSM	RBF (Gui and Kang 2004)	Advanced RBF (Gui and Kang 2004)	The proposed hybrid algorithm
Case 1	Reliability index	2.338	2.330	2.350	2.330	2.332
	Iterations	-	6	4	6	4
Case 2	Reliability index	3.806	3.795	3.799	3.798	3.801
	Iterations	-	6	5	9	4

TABLE 2.2

Analytical Reliability Index of the Brotonne Cable-Stayed Bridge

Failure mode	MCS	FOSM	RSM	Zhang and Liu (2001)	Zhang et al. (2008)	The proposed algorithm
Static bending of girders	3.6104	3.5658	3.5740	3.6193	3.6102	3.6153
Static torsion of girders	-	6.0027	6.0027	6.6027	6.6027	6.0027
Transversal bulking	-	9.7007	9.7397	9.7032	9.7042	9.7126
Longitudinal strength failure	3.5128	3.4946	3.4973	3.5063	3.5146	3.5117

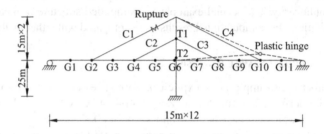

FIGURE 2.4 Dimensions and failure modes of the Brotonne cable-stayed bridge.

(1992). Shen and Wang (1996) provided the limit state functions and distributions of random variables.

As shown in Table 2.2, the number of sampling for MCS is 10^6, which can be treated to get the exact value. The reliability index of the hybrid algorithm has a higher accuracy compared to the FOSM and the RSM. In addition, the analytical result of the hybrid algorithm has a higher accuracy compared to the method provided by Zhang et al. (2008) and Zhang and Liu (2001). One of the benefits of the hybrid algorithm is that the performance function and corresponding design points can be provided, which are not listed in the table.

2.5 CASE STUDY

2.5.1 BACKGROUNDS OF THE SECOND HEJIANG YANGTZE RIVER BRIDGE

The Second Hejiang Yangtze River Bridge is part of the Luyu highways located in Sichuan Province of China. The Kangbo Bridge is a prestressed concrete cable-stayed bridge with main girders of 420 m. The general arrangement diagram of the bridge is shown in Figure 2.5. The material of the main girders is C60 concrete, and the material of the towers is C50 concrete. In total, there are 34 pairs of cables for each pylon supporting 34 pairs of girders. The width of the bridge deck is 30 m, which is divided into three traffic lanes for each driving direction.

All the components of the bridge were numbered in a sequence as shown in Figure 2.5. The girders were marked as GBA1 to GBA34 for the side-span, and GBJ1 to GBA34 for the mid-span. The cables were marked as CBA1 to CBA34 for the side-span and CBJ1 to CBJ34 for the mid-span. The critical sections of the pylon were marked as T1, T2, and T3.

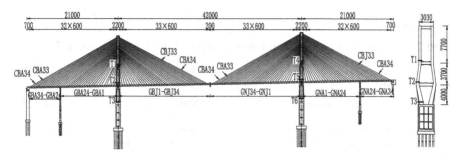

FIGURE 2.5 General arrangement diagram of the Second Hejiang Yangtze River Bridge (unit: cm).

2.5.2 RANDOM VARIABLES

In general, the random variables of a bridge are associated with the materials and external loads. Therefore, the selected random variables in the present study are the elasticity modulus E_i of the main girder, towers, and stay cables, the cross-sectional area A_i, the bending inertia moment I_i, the material density γ_i, the secondary dead load q_s, and the vehicle load q_k. The statistical parameters of these basic random variables were referred to for the empirical distribution.

A bridge suffers from many types of loads. In addition, the loading positions are complex under normal circumstances. Structural reliability analysis is primarily associated with the dead load and live load for simplification. The live load on the deck was assumed as a uniform distribution to simplify the calculation. The structural static reliability is lower when considering a uniformly distributed load on the mid-span for the main girder (Biondini et al. 2008). Therefore, the vehicle load was simplified as a uniformly distributed load on the mid-span girders. Table 2.3 summarizes the statistical parameters of the random variables.

2.5.3 LIMIT STATE FUNCTIONS

Long-span cable-stayed bridges are statically indeterminate due to the complex elements including main girders, pylons, and stay cables. The structural performance function has a higher degree of nonlinearity, as a result of which it is difficult to determine the major failure mode. The serviceability limit state is defined as the event which occurs when a structural stress or a crack exceeds a threshold. Hence, two failure modes are considered as the serviceability limit states: rupture of a stay cable and maximum displacement of the girders.

The limit state functions are shown in Equations (2.1) and (2.2). The serial numbers of the critical components are CBA1-CBA34, CBJ1-CBJ34, CNJ1-CNJ34, and CNA1-CNA34. The strength failure event was caused by the longest four stay cables including CBA34, CBJ34, CNJ34, and CNA34 on the north tower. The yield strength σ_b of the cables is 1860 MPa. The threshold of the vertical deflection u_{max} for the main girders of the cable-stayed concrete bridge is 0.84 m under the vehicle load without considering the dynamic load effect.

TABLE 2.3

Statistical Parameters of the Random Variables of the Cable-Stayed Bridge

Random variables	Symbol	Distribution shape	Mean value	Standard variance
Elasticity modulus of the main girder (kN·m^{-2})	E_1	Normal	3.64×10^7	3.64×10^6
Elasticity modulus of the tower (kN·m^{-2})	E_2	Normal	3.52×10^7	3.52×10^6
Elasticity modulus of the stay cables (kN·m^{-2})	E_3	Normal	1.95×10^8	1.95×10^7
Cross-sectional area of the main girder (standard beam section) m^2	A_1	Lognormal	20.846	1.042
Cross-sectional area of the main girder (auxiliary pier) m^2	A_2	Lognormal	24.694	1.235
Cross-sectional area of the bottom pylon	A_3	Lognormal	26.868	1.343
Cross-sectional area of the top pylon	A_4	Lognormal	22.280	1.114
Cross-sectional area of a single strand in stay cables	A_5	Lognormal	1.4E-4	7.0E-6
Equivalent density of the main girder (kN·m^{-3})	γ_1	Normal	26.56	1.33
Equivalent density of the pylon (kN·m^{-3})	γ_2	Normal	26.24	1.31
Equivalent density of the stay cable (kN·m^{-3})	γ_3	Normal	78.5	3.93
Cross-sectional inertia moment of the main girder (standard beam section)	I_1	Lognormal	18.598	0.930
Cross-sectional inertia moment of the main girder (auxiliary pier)	I_2	Lognormal	23.015	1.151
Cross-sectional inertia moment of the top pylon	I_3	Lognormal	120.864	6.043
Cross-sectional inertia moment of the bottom pylon	I_4	Lognormal	275.517	13.776
Dead load of the bridge deck (kN·m^{-1})	q_s	Normal	132	6.6
Vehicle loading on the mid-span girder (kN·m^{-1})	q_k	Gumbel	63.5	6.35

2.5.4 Reliability Analysis Based on Intelligent Neural Networks

The structural behavior of a long-span cable-stayed bridge is impacted by the nonlinearity. A previous study (Han 2011) indicated that linear analytical results are not conservative compared with the nonlinear analytical result. This study utilized the proposed hybrid algorithm integrating the RSM, FEM, GA and importance MCS methods to analyze the static reliability of the main girder for the Second Hejiang Yangtze River Bridge.

Initially, the finite element model of the cable-stayed bridge was established via a commercial package Midas Civil as shown in Figure 2.6. The geometric nonlinear effects of the sag effect of stay cables was considered via the equivalent elastic modulus method. The main girder and towers were simulated by 470 beam elements, and the stay cables simulated by tension-only truss elements. The cable forces were updated to be consistent with the measured forces. The effects from the prestress

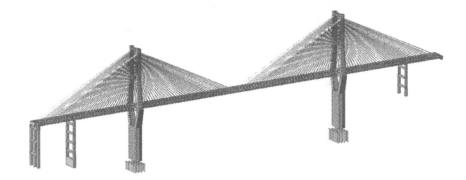

FIGURE 2.6 Finite element model of the cable-stayed bridge.

tendons, the concrete shrinkage, and creep were not considered in the finite element model.

An RBF neural network was trained to approximate the structural static mechanical behavior. The input layer of the network is composed of 17 nodes, and the output layer composed of five nodes corresponding to the displacements of the mid-span girders and cable forces of CBA34, CBJ34, CNJ34, and CNA34. The parameters of the RBF neural network are summarized in Table 2.4. Comparisons of the bridge displacements evaluated from the RBF networks and the FEA are shown in Figure 2.7.

The vehicle load was considered as uniformly distributed in the mid-span of the main girder for the serviceability limit state. The structural reliability was evaluated based on the proposed hybrid algorithm. The corresponding static reliability β of the displacement criterion for the mid-span girder is 7.394. The reliability indices of the strength failure for the #CBA34, #CBJ34, #CNJ34, and #CNA34 cables are $\beta_{CBA34} = 4.973$, $\beta_{CBJ34} = 5.236$, $\beta_{CNJ34} = 5.112$, and $\beta_{CNA34} = 4.867$, respectively. The five reliability indices are shown in Figure 2.8.

2.5.5 Parametric Sensitivity Analysis

The structural reliability index β is affected by the mean value and standard deviation of the random variables. Therefore, this study investigated the influence of the mean value and standard deviation of the random variables on the reliability index. The

TABLE 2.4
Parameters of the RBF Neural Network

Items	Values	Notes
Nodes of the input layer	17	Random variables in Table 2.3
Nodes of the output layer	5	u_{mid} and T_{cab}^i (i = CBA34, CBJ34, CNJ34, CNA34)
Number of training data	200	Uniform design scheme ($U_{200}(200^{17})$)
Number of test data	30	Random selection
Training accuracy	1e-6	With normalization
Testing accuracy	1%	Without normalization

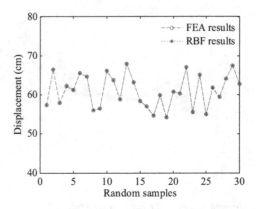

FIGURE 2.7 Comparison of the bridge displacements evaluated from the RBF networks and the FEA.

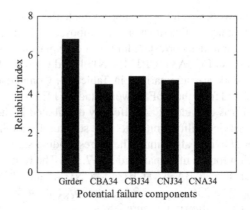

FIGURE 2.8 Reliability indices of the critical failure components.

sensitivities of the reliability index β to the mean value μ_i and standard deviation σ_i of the random variables were evaluated by the differential equations (Hohenbichler and Rackwitz 1986)

$$
\begin{cases}
\dfrac{\partial \beta}{\partial \mu_i} \approx -\alpha_i \\[2mm]
\dfrac{\partial \beta}{\partial \sigma_i} = -\beta \alpha_i^{2}
\end{cases}
\tag{2.6}
$$

where α_i is the direction of the ith random variable in the standard normal distribution space, β is the reliability index, and μ_i and σ_i are the mean value and standard deviation of the ith random variable, respectively. The sensitivity of the reliability index for different means and standard deviations of a random variable was observed under

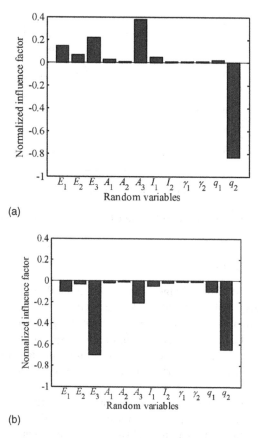

(a)

(b)

FIGURE 2.9 Sensitivity of the reliability index to the means and standard deviations of the random variables: (a) Mean values of the random variables; (b) Standard deviations of the random variables.

a vehicle load distributed throughout the entire main span at a mid-span point of the main span, as shown in Figure 2.9. The figure illustrates that the reliability index is most sensitive to the mean value and standard deviation of the vehicle load q_k.

The variation tendencies of the reliability index for the displacement criterion of the mid-span of the main girder and the strength failure state of the four longest stay cables are shown in Figure 2.10, where the mean coefficient for the vertical load q_k was considered between 1.0 and 2.0.

As observed from Figure 2.10, the reliability index decreases with an increase of coefficient in the mean value. In addition, the reliability index of the displacement criterion is higher than that of the stay cables. The displacement criterion reliability decreases by 30.78% accounting for the mean coefficient increasing from 1.0 to 2.0. The reliability for the cable strength failure decreases by 16.55% to 18.92%. The decreasing trend becomes more serious with the continuous increase of the mean coefficient.

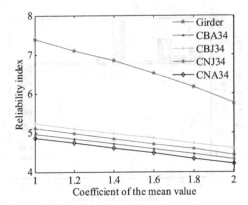

FIGURE 2.10   Reliability index of critical failure locations under different mean coefficients of vehicle loads.

2.6   CONCLUSIONS

This chapter presented a hybrid algorithm for the reliability analysis of long-span bridges via intelligent RBF neural networks. Feasibility of the hybrid algorithm was demonstrated by two numerical examples. The hybrid algorithm method was subsequently utilized for the reliability evaluation of a long-span PC cable-stayed bridge under a vehicle load. The influence of random variables on the structural reliability index was indicated in the parametric sensitivity analysis. The conclusions are summarized as follows.

The hybrid approach organically integrates the RBF neural network, the GA, and the MCS method, and thus can provide an accurate and efficient reliability evaluation. The analysis results of two numerical studies indicated that the hybrid approach has a satisfactory accuracy and calculation efficiency. The reliability index for the strength failure of stay cables is more critical for that of the displacement criterion of the mid-span girder. The parametric sensitivity analysis indicated that the reliability index of a cable-stayed bridge is sensitive to the mean and standard deviation of the vehicle load. The other sensitive factors are the cross-section area of the stay cable and the mean and standard deviations of the elasticity modulus. The reliability indices of the displacement criterion and strength failure of the cable decrease proportionately to the increase of the vehicle load. Furthermore, the decreasing rate becomes more serious with the continuous increase of the mean coefficient.

REFERENCES

An, X., Gosling, P. and Zhou, X. 2016. Analytical structural reliability analysis of a suspended cable, *Structural Safety* 58: 20–30.

Arangio, S., and Bontempi, F. 2015. Structural health monitoring of a cable-stayed bridge with Bayesian neural networks, *Structure and Infrastructure Engineering* 11(4): 575–587.

Biondini, F., Frangopol, D.M. and Malerba, P.G. (2008), Uncertainty effects on lifetime structural performance of cable-stayed bridges, *Probabilistic Engineering Mechanics*, 23(4), 509–522.

Bruneau, M. 1992. Evaluation of system-reliability methods for cable-stayed bridge design, *Journal of Structural Engineering* 118: 1106–1120.

Catbas, F. N., Susoy, M., and Frangopol, D. M. 2008. Structural health monitoring and reliability estimation: long span truss bridge application with environmental monitoring data, *Engineering Structures* 30(9): 2347–2359.

Chehade, F. and Younes, R. 2020. Structural reliability software and calculation tools: a review, *Innovative Infrastructure Solutions* 5(1): 1–10.

Chen, T., Wang, S. and Shi, Z. 2000, Reliability analysis of cable-stayed bridges considering geometrical non-linearity, *Journal Tongji University (Natural Science)* 28(4): 407–412. (in Chinese).

Cheng, J. and Li, Q.S. (2009), Reliability analysis of a long span steel arch bridge against wind-induced stability failure during construction, *Journal of Constructional Steel Research*, 65(3): 552–558.

Cheng, J., Li, Q.S. and Xiao, R.C. 2008. A new artificial neural network based response surface method for structural reliability analysis, *Probabilistic Engineering Mechanics* 23(1): 51–63.

Cheng, J. and Xiao, R. 2004. Static reliability analysis of cable-stayed bridge, *Journal Tongji University (Natural Science)* 32(12): 1593–1598.

Deng, Y., Li, A. and Feng, D. (2018). Fatigue reliability assessment for orthotropic steel decks based on long-term strain monitoring. *Sensors* 18(1), 181.

Frangopol, D. and Imai, K. 2004. Reliability of long span bridges based on design experience with the Honshu–shikoku bridges, *Journal of Constructional Steel Research* 60(3/5): 373–392.

Gong, Q., Zhang, J., Tan, C. and Wang, C. 2012. Neural networks combined with importance sampling techniques for reliability evaluation of explosive initiating device, *China Journal of Aeronautics Chinese* 25(2): 208–215.

Gui, J. and Kang, H. 2004. The study of the whole response surface method for structural reliability analysis, *Journal of Building Structures* 25(4): 100–105.

Han, S. H. 2011. A study on safety assessment of cable-stayed bridges based on stochastic finite element analysis and reliability analysis, *KSCE Journal of Civil Engineering* 15(2): 305–315.

Hohenbichler, M. and Rackwitz, R. (1986), Sensitivity and importance measures in structural reliability, *Civil Engineering System* 3(4): 203–209.

Kong, X., Wu, D.J., Cai, C.S. and Liu, Y. Q. 2012. New strategy of substructure method to model long-span hybrid cable-stayed bridges under vehicle-induced vibration, *Engineering Structures* 34: 421–435.

Li, X.J. and Hu, Y.C. (2000), Study on the approximate structural reliability analysis method based on GA-NN combined technology, *China Civil Engineering Journal* 33(5): 40–45.

Liu, Y., Lu, N., Yin, X., and Noori, M. 2016. An adaptive support vector regression method for structural system reliability assessment and its application to a cable-stayed bridge, *Proceedings of the Institution of Mechanical Engineers, Part O: Journal of Risk and Reliability* 230(2): 204–219.

Shen, H. and Wang, Y. 1996. Static and dynamic analysis of girders of cable-stayed bridges, *China Journal of Highway Transportation* 9(1): 59–65.

Tang, Q. and Zhang, C. 2013. Data Processing System (DPS) software with experimental design, statistical analysis and data mining developed for use in entomological research, *Insect Science* 20(2): 254–260.

Wang, J., Zhang, J., Xu, R., and Yang, Z. (2019). Evaluation of thermal effects on cable forces of a long-span prestressed concrete cable–stayed bridge, *Journal of Performance of Constructed Facilities* 33(6): 04019072.

Wu, F. and Zhao, L. 2006. Static reliability analysis of the main girder of concrete cable-stayed bridge, *Journal of China Railway Science* 28(3): 88–91.

Yan, D. and Chang, C. 2009. Vulnerability assessment of cable-stayed bridges in probabilistic domain, *Journal of Bridge Engineering* 14(4): 270–278.

Yang, D., Yang, L., and Chen S. 2017. Long-term in-service monitoring and performance assessment of the main cables of long-span suspension bridges, *Sensors* 17(6): 1414.

Zhang, Q., Bu, Y. and Li, Q. 2008. Hybrid chaos algorithm of structural reliability analysis of Cable-stayed bridge, *China Journal of Highway Transportation* 21(3): 48–52.

Zhang, J.R. and Liu, Y. 2001. Reliability analysis of cable-stayed bridge using GAS and ANN, *China Civil Engineering Journal* 34(1): 7–13 (in Chinese).

3 System Reliability Assessment of a Cable-stayed Bridge Using an Adaptive Support Vector Regression Method

Yang Liu

Hunan University of Technology, China

Naiwei Lu

Changsha University of Science and Technology, China

Mohammad Noori

California Polytechnic State University, USA

CONTENTS

3.1 INTRODUCTION

Engineering structures involve uncertainties associated with initial imperfections in the structural systems, resistance deterioration, and increasing loads. Thus, probabilistic safety assessment of engineering structures has become increasingly important in recent decades. Structural reliability is a probabilistic method that takes uncertainties in structural analysis into consideration. In particular, fatigue reliability, dynamic reliability and seismic reliability assessment of engineering structures have been widely studied in the recent years (Casciati et al. 2008; Zhang et al. 2012; Baldomir et al. 2013). Since engineering structures are mostly statically indeterminate structures consisting of various types of components, the failure modes are random in nature. A single failure of one component may not lead to failure of the entire structure. For example, a cable-stayed bridge is composed of stay cables, girders, and towers. The potential failure modes are bending failure of girders, strength failure of cables, and displacement failure of girders. The failure of a cable may not lead to failure of the entire bridge. Thus, system reliability evaluation of the engineering structure deserves investigation.

For system reliability assessment of these structures, two key issues need special consideration. Foremost, the structural performance functions are implicit with high order nonlinear behaviors. The commonly used methods associated with Artificial Neural Network (ANN), Respond Surface Method (RSM), and Monte Carlo Simulation (MCS) have been developed to address these problems. For nonlinear performance functions, Chan and Low (2012) demonstrated that the neural network approach which did not involve any predetermined model had advantage over the response surface approach which was usually based on a low-order polynomial assumption. Since the principle in ANN is based on the least mean square error (MSE), local convergence and over fitting are the key problems to be overcome when they are used to approximate the implicit functions. In order to address these problems, a support vector machine (SVM) with the principle of structural risk minimization (SRM) was proposed as an alternative approach. Tan et al. (2011) proposed a radial basis function neural network-based SVM method to reduce computational costs in structural reliability analysis. Lins et al. (2012) utilized the SVM and the particle swarm optimization (PSO) approach to conduct reliability analysis based on time series data of engineered components. Wei et al. (2013) proposed a dynamic SVR method to solve reliability prediction problems. Khatibinia et al. (2013) proposed a wavelet weighted least square SVM method to reduce the computational cost in seismic reliability assessment of existing reinforced concrete structures. However, the scope of the application and utilization of SVM method in structural system reliability assessment is relatively deficient.

Another main challenge to system reliability analysis is that there are numerous possible failure sequences. Without an efficient search scheme, an excessive number of failure sequences are necessary in order to propose an accurate estimate of the failure's probability. Furthermore, quantifying the likelihood of system-level failure sequences requires drastically new structural analyses in order to account for the load redistributions and various uncertainties. Thus, an overwhelming computational cost is required in structural system reliability assessment, and the MCS method is the most straightforward approach among the existing analysis methods. The computational process of MCS method can be exhaustive when computational simulations are time consuming or when the failure probabilities are low (Liang et al. 2011). Therefore, several non-sampling-methods have been developed to increase the computational efficiency of the transitional MCS method. The widely used non-sampling method is selective, searching schemes based on an event tree of potential failure sequences, such as the branch-and-bound (B&B) method developed by Murotsu (1984). In order to overcome the time-consuming problem and the possible misidentification of the critical failure sequences in the B&B method, Lee and Song (2012) developed a new branch-and-bound (B3) method for FE-based system reliability analysis of continuum structures by modifying the LSFs. Kang et al.(2008) proposed a matrix-based system reliability (MSR) method that generalizes the linear programming bounds method to represent the system events in the case of complete information. Kim et al. (2013) proposed a simulation-based selective searching order using a genetic algorithm and the system reliability was computed based on a multi-scale MSR method. Although extensive efforts have been developed to structural system reliability assessment, the trade-off between time-consuming computation and efficiency is still a challenging issue.

In order to address these two challenges, in this chapter a new adaptive support vector regression (ASVR) method for structural system reliability assessment is proposed. Compared with traditional SVR, the proposed ASVR method involves two updating procedures of support vectors to approximate structural LSFs. The first procedure involves approximating the LSF at the most probable failure point (MPFP) based on genetic algorithm, and subsequently importance sampling MCS can be used to estimate structural reliability with a reasonable accuracy. The second procedure includes approximating LSFs once any component has failed, and then the failure sequences are established based on the β-bound method. With the two updating processes, the traditional SVR is improved and applied to engineering structural system reliability assessment. Two verification examples involving a nonlinear LSF and a geometrically nonlinear elastic truss are presented in order to illustrate the accuracy and efficiency of the proposed method. Finally, a pre-stressed concrete cable-stayed bridge with a mid-span of 420 m is studied to demonstrate applicability of the proposed method.

3.2 SUPPORT VECTOR REGRESSION METHODOLOGY FOR STRUCTURAL RELIABILITY ESTIMATION

The purpose of structural reliability estimation is to obtain the probability of failure P_f involved with uncertain design parameters (e.g., material parameters and loadings). The value of P_f can be mathematically expressed by the multivariate integration

$$P_f = \int_{g(x \leq 0)} f(x)dx \tag{3.1}$$

where, x is the random variable vector, $g(x)$ and $f(x)$ represent the LSF and joint probability density function, respectively.

For most engineering structures, the boundary condition $g(x) = 0$ exhibits a highly nonlinear and implicit function. Furthermore, the conventional numerical integration methods are not applicable to these structures. Although as discussed above many good approaches have been proposed to approximate the LSFs, how to determine the trade-off between the sampling number and the calculation precision remains a challenge. If the sampling number is insufficient, the solution may not be credible; on the other hand, if the sampling number is increased so too will the computational time increase.

In order to increase computational efficiency and ensure precision of the approximated responses, in this paper the SVR method is adopted and developed to replace the actual LSFs. The traditional SVR method for structural reliability estimation contains two main steps: approximating the structural responses and estimating the reliability. The methodology will be illustrated in the following section.

3.2.1 APPROXIMATING THE STRUCTURAL RESPONSES BASED ON AN SVR METHOD

The SVM is a machine learning scheme that is becoming increasingly popular in structural reliability assessment. Vapnik (2000) first proposed SVM and applied it in support vector machine for classification (SVC) and SVR. The SVR theory used in this study is briefly described as follows.

The structural LSFs can be approximated by an SVR model, based on underlying concepts that arise from statistical learning theory. The learning error is measured by means of loss ε-insensitive function which is defined as

$$e(x,y,g) = \max\left(0, \left|y - g(x) - e\right|\right) \tag{3.2}$$

where $g(x)$ is the predicted value, y is the observed value, e is an error-free tube of radius. The ε-insensitive principle shown in Figure 3.1 is the principle to check whether the difference between the predicted value $g(x)$ and the observed value y is larger than ε.

In Figure 3.1, ζ and ζ^* denote slack variables that measure the error of down and up sides, respectively. The ε-insensitive principle indicates that increasing ε decreases the corresponding ζ and ζ^*, thereby reducing the error resulting from the corresponding sampling points.

For actual engineering structures, the relationship between random variables x and structural responses $g(x)$ are nonlinear. Thus, a linear regression with a

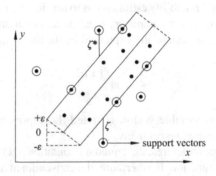

FIGURE 3.1 ε-insensitive principle (Suykens and Vandewalle 1999).

high-dimension feature space shown in Equation (3.3) is employed to approximate the responses.

$$g(x) = w \cdot \varphi(x) + b \qquad (3.3)$$

In Equation (3.3), x denotes the random variable vector, $\varphi()$ is a kernel function, w denotes the weight vector in primal weight space, and b is the bias term. An optimization method can be adopted to obtain these parameter values. With the SRM principle, the SVR is formulated as minimization of the following function

$$\min J(w,e) = \frac{1}{2}\|w^2\| + \frac{1}{2}C\sum_{i=1}^{L}e_i^2 \qquad (3.4)$$
$$\text{s.t. } y_i = w^T\varphi(x_i) + b + e_i \, (i = 1,2,\ldots,L)$$

where C is a regularization constant, and e is an error vector. In order to solve the optimum problem shown in Equation (3.4), Suykens and Vandewalle (1999) introduced the Lagrange multiplier optimal programming method described as

$$L(w,b,e,\alpha) = J(w,e) - \sum_{i=1}^{l}\alpha_i\left(w^T\varphi(x_i) + b + e_i - y_i\right) \qquad (3.5)$$

where α_i is the Lagrange multipliers, which can be either positive or negative due to the Karush-Kuhn-Tucker conditions (Moura et al. 2011). More details regarding the description of the solution process can be found in Deng and Yeh (2011). According to Equations (3.3–3.5), the resulting least squares SVR model for function estimation is represented as

$$f(x) = \sum_{i=1}^{l}\alpha_i\psi(x,x_i) + b \qquad (3.6)$$

where $\psi(x,x_i)$ is the kernel function which enables the dot product to be computed in a high-dimension feature space using low-dimension space data input without the transfer functions. The kernel function in this paper is a Gaussian function. Compared with the traditional SVR, least squares SVR exhibits an efficient learning rate, since the inequality constraints are replaced with equality constraints (Liu and Zhu 2013).

In order to build an accurate and efficient SVR model, the selection of training sample datasets and the parameter of the SVR model should be carefully specified. Regarding the training sample detests, there are many excellent design schemes, such as the orthogonal design and the uniform design. The latter method accommodates the largest possible number of levels for each factor among all experimental designs and could be introduced to make the selected training data as uniform as possible to cover the entire design space. It has been applied in the fields of chemistry and other engineering topics, but its application in reliability analysis is limited. The most important part of using uniform design method in structural reliability analysis is described as follows (Cheng et al. 2007). First of all, the range of domain for the variables should be determined by the three-sigma rule or empirical rule. Next, a uniform design table should be selected and denoted by $Un(q^t)$, where "U" represents the uniform design table, n is the number of experiments, q is the number of level for each factor, and t is the number of variables. At the end, the samples are set by numbers associated with the uniform design table.

Regarding the parameters of SVR model, the kernel function $\psi(x,x_i)$, regularization constant C, bandwidth of the kernel function σ, and tube size of ε-insensitive loss function should be set by trial-and-error procedure or an optimization method (Cheng and Li 2008). Herein, the structural responses can be efficiently approximated based on the SVR approach.

3.2.2 ESTIMATING STRUCTURAL RELIABILITY USING THE SVR-MCS APPROACH

Once the explicit LSFs (substituted by SVR) are obtained, the P_f in Equation (3.1) can easily be estimated by traditional reliability method (e.g., FORM and MCS). In this paper, an importance sampling method (the MCS) was used to estimate the reliability at the MPFP. The traditional MCS method allows the determination of an estimate of the probability of failure, given by (Wu et al. 2009)

$$\bar{p}_f = \frac{1}{N}\sum_{i=1}^{N} I(A) \tag{3.7}$$

where N is the total sampling number, and $I(A)$ is an indicator function for event A defined by

$$I(A) = \begin{cases} 1 \text{ if } g(X) < 0 \\ 0 \text{ if } g(X) \geq 0 \end{cases} \tag{3.8}$$

where the approximate value $g(X)$ is obtained by the ready-made SVR surrogate model illustrated beforehand. In order to guarantee the availability of the failure probability, the total sampling number N should satisfy the requirement:

$$N \geq \frac{100}{\Phi(-\beta_0)} \times \frac{\Phi(-\beta_0) - \Phi^2(-\beta_0)}{\Phi(-2\beta_0)\exp(\beta_0^2) - \Phi^2(-\beta_0)} \tag{3.9}$$

where β_0 is the estimated reliability index. For most cases of practical interest, N is extremely large. Consequently, variance reduction techniques have been performed. The most popular method is to use the importance sampling technique. In order to carry out simulation with samples having a higher rate of falling in the failure region, Equation (3.7) can be rewritten as (Cardoso et al. 2008):

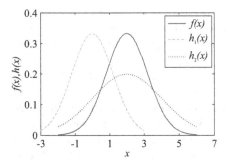

FIGURE 3.2 Relationship for $f(x)$, $h_1(x)$ and $h_1(x)$.

$$\overline{P}_f = \frac{1}{M} \sum_{i=1}^{M} I\left[g(v_i) \leq 0\right] \frac{f_X(v_i)}{h(v_i)} \qquad (3.10)$$

where M is the total sampling number, $f_X()$ and $h()$ are the actual probability density function (PDF) and importance sampling PDF, respectively, and v_i are the samples distributed according to $h()$. Two importance sampling PDF functions for a simple random variable with single failure mode are plotted in Figure 3.2.

With the two importance sampling functions, shown in Figure 3.2, more sampling points will fall into the failure region, which provides a significant contribution to the calculation efficiency. The variance of the estimator $\overline{P}_f$ could be reduced if the importance sampling density function $h()$ is selected properly. Kernel density-based method (Dai et al. 2012a) and the design point method (Botev et al. 2013) are both well-developed importance sampling methods. The key issue for a design point method is to find the MPFP from the LSF. The idea for solving the key issue will be illustrated below.

3.3 THE PROPOSED ASVR METHOD FOR STRUCTURAL SYSTEM RELIABILITY

In the previous section, SVR-based reliability estimation methodology was examined. The use of SVR is to approximate the LSFs, which may reduce the total effort on the reliability evaluation, compared with the traditional MCS and FOSM method. Thus, the SVM-based reliability estimation method is widely used and improved for structural reliability assessment. Numerous novel and improved algorithms have been proposed (e.g., radial basis function neural network-based RSM approach, SVM-based RSM approach, and genetic algorithm-based SVM approach). Since reliability assessment of actual structures involves significant computational efforts, the application of SVR-based reliability evaluation method is still limited. On the other hand, limited investigation has been performed for SVM-based structural system reliability estimation. The most important step in structural system reliability analysis is to identify structural failure sequences. The traditional SVR-based method cannot be directly applied to identify structural failure sequences. Accordingly, additional work should be done to improve and apply SVR method to structural system reliability evaluation.

The customary method of establishing structural failure sequences is the β-bound method (Botev et al. 2013; Liu et al. 2014). According to this method, the components with more failure probability are deleted, and then incorporated to the subsequent screen procedure. When the components have failed, the structural stiffness is changed according to the failure sequences. Thus, once any single component fails, structural response functions should be updated too. According to the aforementioned ideas, an ASVR method is proposed in this paper to solve this problem. The framework of the proposed ASVR is shown in Figure 3.3.

In Figure 3.3, the "SVR model" procedure is similar to the traditional SVR modeling step, which can be found in the literature (Vapnik 2000; Suykens and Vandewalle, 1999; Liu et al., 2016). There are three main steps in the framework: "SVR-MCS-based reliability estimation," "the first updating procedure: searching the MPFP," and "the second updating procedure: searching failure sequences." Compared with the traditional SVR method, the proposed ASVR method includes two updating processes. The updating procedures are discussed in this section.

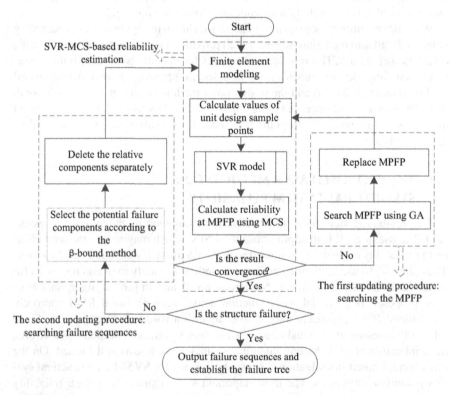

FIGURE 3.3 Framework of the proposed ASVR-based system reliability estimation method.

3.3.1 THE UPDATING IDEA

The idea of updating procedure in ASVR framework was initially introduced by Vanli and Jung (2014) and Degrauwe, et al. (2009), where a finite element model updating procedure was used to investigate damage assessment of structures. In order to update damage assessment models, this study utilized an additive bias correction model.

$$y^e(x) = y^m(x,\theta) + \delta(x) + \varepsilon \tag{3.11}$$

where x is the input variable vector, $y^e(x)$ is the response measured experimentally, $y^m(x)$ is the model prediction of the response, ε is the random experimental error vector, and $\delta(x)$ is a bias correction function which is assumed to be a Gaussian process. The purpose of model updating is to find the parameters of the bias correction function that provide the best agreement between y^m and y^e.

However, the purpose of the updating procedures in this paper is to update support vectors in the SVR model. By combining Equation (3.6) with Equation (3.11), the formulation of updating the SVR model can be derived as

$$g(x) = w \cdot \varphi(x) + b \approx g(x,\theta) + \delta(x) + \varepsilon \tag{3.12}$$

where w and b denote the weight vector in primal weight space and the bias term, respectively; w and b are calibration parameters that were updated in the SVR model. Once the relative components fail according to the β-bound method, the structural finite element model will be modified and the ready-made SVR surrogate model will be updated too.

3.3.2 THE FIRST UPDATING PROCEDURE: SEARCHING THE MPFP

As shown in Equation (3.10), importance sampling density function $h()$ is associated with the MPFP. Thus, the first updating procedure is searching the MPFP based on the optimization problem described as

$$\begin{cases} \min \ \beta^2 = \sum_{i=1}^{n} \left[\left(x_i - \mu'_{x_i} \right) / \sigma'_{x_i} \right]^2 \\ \text{s.t. } g(x_1, x_2, \cdots x_n) = 0 \end{cases} \tag{3.13}$$

where β is the minimum distance from the origin to the LSF $g(X)$ in the so-obtained standard normal space, x_i is the ith variable that needs to be optimized, μ'_{x_i} and σ'_{x_i} are the mean value and standard deviation of the ith equivalent independent normal random variable, respectively.

The conventional method to solve Equation (3.13) is the Hasofer Lind and Rackwitz Fiessler (HOF) method which involves constrained optimization and iteration (Du and Hu 2012). Dai et al. (2012b) used the Markov chain simulation to generate samples in the failure region, and then used SVR to obtain the explicit approximation of the LSF. Compared with Dai's method, our paper utilized IS together with GA to search the MPP and the calculation reliability index. In order to increase the computational efficiency, a genetic algorithm was employed in this study to search the MPFP (Equation (3.13)).

The genetic algorithm generates solutions to optimization problems using techniques inspired by natural evolution, such as inheritance, mutation, selection, and crossover. In each generation, the fitness of every individual in the population is evaluated by objective function. There are two advantages for genetic algorithm used in solving the MPFP of the LSF. First, the explicit LSF was approximated using SVR approach from which the 's.t.' function in Equation (3.13) could be quickly calculated. Second, the genetic algorithm involves a penalty function which translates Equation (3.13) into an unconstrained optimization problem. A penalty function is used in this paper and the general form is

$$\begin{cases} m = \alpha \times (1 - gen \,/\, maxgen) + b \\ \beta' = \beta + m \cdot \text{abs} \big[g(\mathbf{x}) \big] \end{cases} \tag{3.14}$$

where m is the penalty coefficient which decreases gradually, α is a coefficient of punishment, b is a penalty parameter greater than 0 which assures the validity of punishment, gen is the present iteration, $maxgen$ is the maximum value of iteration, abs() is the absolute function to assure the efficiency of the punishment, β' is the penalized objective function, β is the primary value in Equation (3.13). α and b can be obtained by pilot calculation.

Since the original LSFs are approximated in the vicinity of the mean values of basic random variables, the response value $g(X^*)$ at the optimized MPFP X^* is different from the actual structural response. To improve the accuracy of the LSF value, an alternative procedure of selecting the sampling points (Bucher and Bourgund 1990) was utilized. The function to generate the new center point from linear interpolation is

$$x_M^{(k)} = x^{(k)} + \left(x^{*(k)} - x^{(k)} \right) \frac{g\left(x^{(k)} \right)}{g\left(x^{(k)} \right) - g\left(x^{*(k)} \right)} \tag{3.15}$$

where k corresponds to the kth iteration, x_M is the new center point, x and x^* are the mean value point and optimized point, respectively, and $g(x)$ and $g(x^*)$ are the function values of x and x^* in the SVR-based approximate function, respectively. This step guarantees the sampling point x_M chosen from the new center point x^* including the information from the original failure surface sufficiently.

Herein, with some iteration, the MPFP will be obtained and Equation (3.10) can be employed to calculate P_f with a relative high precision.

3.3.3 THE SECOND UPDATING PROCEDURE: SEARCHING FAILURE SEQUENCES

After the first updating procedure, the structural components' reliability is estimated. The following procedure was introduced to search the failure sequences of the structure. The second part of the updating procedure is updating SVR models to identify the failure components and to establish the fault tree, according to the β-bound method.

For the purpose of identifying the failure components, there are two sub-steps in the second updating procedure. The first is to search the failure components according to the β-bound method. Fundamental analytical methods of the β-bound have been presented by several researchers (Liu et al. 2014). In the β-bound method, a structural system failure event E_s is made of m independent failure modes $E_i(i = 1,...,m)$, where E_i includes some failure events $E_i^j(j = 1,...,n)$ in sequences. The structural system reliability is obtained by

$$
\begin{cases}
E_i = \bigcap_{j=1}^{n} E_i^j \\
E_s = \bigcup_{i=1}^{m} E_i
\end{cases}
\tag{3.16}
$$

As shown in Equation (3.16), single failure events are composed of multiple failure states in parallel, and system reliability failure events are composed of multiple failure events in series. Accordingly, the entire structure can be modeled as a series–parallel combination of failure modes. The system reliability is estimated on basis of failure sequence models.

Based on risks or probabilities of failure, the reliability range in β-bound method for identification of critical failure is

$$
\beta_{r_k}^{(k)} = \left[\beta_{min}^{(k)}, \beta_{min}^{(k)} + \Delta\beta^{(k)} \right]
\tag{3.17}
$$

where (k) refers to the kth stage of the failure process, β_{rkl} denotes the contingent reliability index of the r_k component, β_{min} is the minimum reliability index, and $\Delta\beta$ is the reliability index range. $\Delta\beta$ is equal to 3 in the first stage and equal to 1 thereafter.

The second is deleting the relative components separately and updating the SVR model. The structural finite element model is updated in this step, and then enters into the modeling SVR step. From that point onward, the first stage of establishing failure sequences is finished. At the same time, the SVR model is updated and the parameters in Equation (3.12) are also modified. The iterations are taken on with the framework until all the failure sequences are identified.

The benefit of the second procedure is that the updating procedure is computationally efficient. Furthermore, the updating procedures associated with the SVR model enable the excellent SVR approach to be used in a structural fault tree.

3.3.4 CORRELATION COEFFICIENT CALCULATION

When the potential failure elements and failure sequences are found, the next step is calculating correlation coefficients of these elements and sequences. There are two types of correlation coefficient calculation in structural system reliability evaluation: the correlation of random variables and of failure modes.

Regarding the correlation of random variables, the three most representative transformative methods are orthogonal, Rosenblatt, and Nataf transformation. This paper simply introduced the Resonblatt transformation which maps the correlated variables onto the independent standard normal variables. The transformation is defined by the following successive conditioning: (Noh et al. 2009)

$$u_1 = \Phi^{-1}\left[F_{X_1}(x_1) \right]$$
$$u_2 = \Phi^{-1}\left[F_{X_2}(x_2|x_1) \right] \tag{3.18}$$
$$u_n = \Phi^{-1}\left[F_{X_n}(x_n|x_1, x_1, \dots, x_{n-1}) \right]$$

where n is the number of input variables, $F_{Xi}(x_i|x_1, x_1, \dots, x_{i-1})$ represents to the CDF of X_i, and Φ^{-1} represents to the inverse CDF of the standard normal variables. When the joint CDF is known the Rosenblatt transformation is an exact solution. In addition, the result of the Rosenblatt transformation is not affected by the ordering adopted for the variables. Even though the Rosenblatt transformation is exact, it can only be used for limited cases where all input variables are independent or a joint CDF is provided. The estimation of failure probability might be different for an input joint non-normal CDF due to the approximation error of FORM.

An equivalent linearization method developed by Johnson (2013) was employed to calculate correlation coefficient in this paper. There are three main steps in this calculation using the Johnson method.

An equivalent linear LSF should be established at the MPFP. The equation can be expressed as

$$\bar{G}_i(Y) = \alpha_i^T Y^* + \beta_i \tag{3.19}$$

where a_i^T is the unit normal vector of the ith failure element at the MPFP, Y^* is the MPFP, β_i is the reliability index of the ith element. After that, the Johnson method is applied to obtain a synthetic equivalent linear LSF from the multiple failure elements. The synthetic LSF is expressed as

$$\bar{G}_E(Y) = \alpha_E^T Y^* + \beta_e \tag{3.20}$$

where a_E^T is the equivalent unit normal vector of the corresponding failure elements and β_e is a comprehensive reliability index of a failure sequence. After all the failure sequences have been processed using Equation (3.20), the correlation coefficients can be calculated by

$$\rho_{12} = \sum_{i=1}^{n} \left(\alpha_{E1i} \cdot \alpha_{E2i} \right) \tag{3.21}$$

where ρ_{12} respects the correlation coefficient between the first and second failure sequences, α_{E1i} and α_{E2i} respects to the equivalent unit normal vector of the ith element in the first and second failure sequences, respectively. Once the correlation coefficients are obtained, the upper bound and lower bound of system reliability can be evaluated by the narrow bounds method.

EXAMPLE 3.1   A NONLINEAR LIMIT STATE FUNCTION

This example was previously employed by Kang et al. (2010). The LSF is expressed as

$$g(X) = \exp\left[0.4(u_1 + 2) + 6.2\right] - \exp\left[0.3u_2 + 5\right] - 200 \qquad (3.22)$$

where u_1 and u_2 are expected to submit to a standard normal distribution.

The first step is to approximate the LSF in the mean value point. In this step a data process system (DPS) (Tang and Zhang 2013) was used to generate uniform design sampling points. The sampling points by uniform design method and Bucher's method are shown in Table 3.1. The optimum parameters of SVR model are shown in Table 3.2. The second step is to search MPFP in the SRV model by the first updating process. The approximated LSF of Example 3.1 at the mean value point and the MPFP are shown in Figure 3.4 (a) and (b), respectively. The iteration of genetic algorithm-based optimization is plotted in Figure 3.5. The reliability was calculated and compared with the other approaches as shown in Figure 3.6.

Since the ASVR-based reliability analysis is finally calculated by the MCS method, the results calculated by MCS are assumed to be the exact solution. The calculated results are shown in Table 3.3.

As observed in Table 3.3, the failure probability of MCS, FORM, RSM, and ASVR are nearly the same. The values of FORM and RSM are larger than the exact value. This error is not the fault of these methods but caused by the approximated linear LSF instead of the actual nonlinear LSF. Since importance sampling MCS is employed in ASVR, the value of ASVR is close to the exact value. In addition the numerical results demonstrate that the number of iterations and samples needed in ASVR is less than the other methods.

In order to analyze the effect of a different number of samples and sample generation schemes on the performance of the ASVR method, the uniform design and Bucher's methods were compared in Table 3.4. It can be seen that the former method agrees reasonably well with the exact solution and requires fewer training samples when compared with the Bucher's method. In addition, the relative error of the uniform design method and the Bucher's method is around 1.30% and 2.13%, respectively, with 51 training samples. This is mainly because the uniformly distributed performance of the uniform design method is better than the Bucher's method. Thus, the MSE of the approximated LSF by the uniform design method is less than that of Bucher's method.

TABLE 3.1
Design of Sampling Points

	Uniform design points for ASVR model		Bucher's points for RSM model	
Number	u_1	u_2	u_1	u_2
1	3	1.5	-3	0
2	1.5	-3	0	-3
3	3	1.5	3	0
4	-1.5	3	0	-3
5	0	0	0	0

TABLE 3.2
Parameters in the SVR Model

Parameters	C	σ	ε
Value	22.62	0.0055	0.062

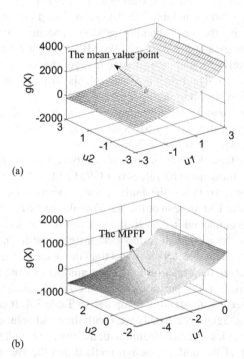

(a)

(b)

FIGURE 3.4 Approximated LSF of Example 3.1: (a) at the mean value point; (b) at the MPFP.

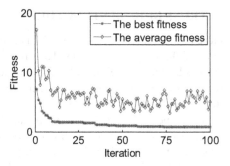

FIGURE 3.5 Iteration of MPFP in the first updating process of Example 3.1.

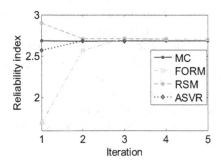

FIGURE 3.6 Comparison of different approaches of Example 3.1.

TABLE 3.3
Comparison of Analytical Results for Example 3.1

Items	Exact solution(MC)	FORM	RSM	SVR	ASVR
Failure probability	0.003605	0.003365	0.003364	0.003924	0.003709
Error (%)	-	−6.66%	−6.68	8.85%	2.88
Number of iterations	-	5	4	-	3
Sample size	10^6	-	20	5	15

TABLE 3.4
Influence of Different Sampling Schemes on the ASVR Method

Sampling scheme	Total number of samples with three iterations	Failure probability	Error/%
Bucher's method	15	0.003741	3.77
	27	0.003694	2.46
	51	0.003682	2.13
Uniform design method	15	0.003709	2.88
	27	0.003674	1.91
	51	0.003652	1.30

EXAMPLE 3.2   A TRUSS BRIDGE STRUCTURE

A truss bridge structure (Kim et al. 2013) in Figure 3.7 is considered to investigate the applicability of the proposed method. The cross-sectional area of each member of the truss bridge is listed in Table 3.5. The applied loads and yield stress of the members listed in Table 3.6 are considered as random variables and assumed to be statistically independent.

The uniform design scheme is $U_{20}(20^2)$, where 20 samples were used to approximate the structural response surface with two variables. The original uniform design points and the updated uniform design points are shown in Figure 3.8. The first candidate failure component was No. 3. In order to demonstrate the second updating procedure, the response surface of component 9 associated with component 3 is shown in Figure 3.8.

There are ten dominant failure modes identified according to the ASVR. These failure modes and corresponding reliability indices are shown in Figure 3.9. In order to examine the accuracy and numerical efficiency of the proposed method, the failure probability of the overall system and corresponding system reliability are compared in Table 3.7.

These results reveal the accuracy and the numerical efficiency of the proposed method. The proposed method was able to identify the domain failure modes of a complex structural system efficiently and compute the system failure probability accurately.

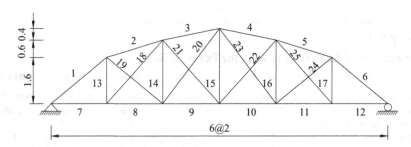

FIGURE 3.7   A truss bridge structure (unit: m).

TABLE 3.5
Cross-Sectional Areas of the Components in Example 3.2

Number of components	Cross-sectional area/m²
1–6	15×10^{-4}
7–12	14×10^{-4}
13–17	12×10^{-4}
18–25	13×10^{-4}

TABLE 3.6

Statistical Parameters of the Random Variables in Example 3.2

Random variables /unit	Distribution	Mean value	c. o. v.
P1/kN	Lognormal	160	0.1
P2/kN	Lognormal	160	0.1
$\sigma y_i, i = 1,...,25$/MPa	Normal	276	0.05

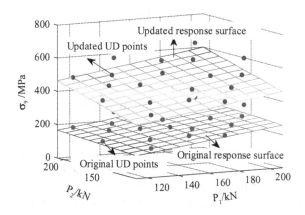

FIGURE 3.8   Response surface of component 9 associated with the component No. 3.

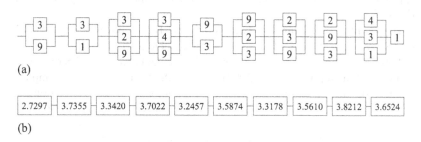

FIGURE 3.9   Dominant failure modes and corresponding reliability indices obtained by ASVR: (a) failure modes; (b) reliability indices.

TABLE 3.7

Comparisons of the Results for the ASVR with Other Methods

Items	MCS(c.o.v. = 0.02)	Kim et al. 2013	ASVR
System reliability index	2.5733	2.5478	2.5212
Failure probability	0.0050367	0.0054202	0.0058477
Error/%	-	7.614	6.498
Number of simulations	460,330	51,344	1200

EXAMPLE 3.3   A GEOMETRICALLY NONLINEAR ELASTIC TRUSS

A geometrically nonlinear elastic truss and hinge elements, as shown in Figure 3.10, was considered for system reliability evaluation (Frangopol and Imai 2000). The statics of random variable parameters are shown in Table 3.8 and Table 3.9.

All variables in Tables 3.8 and Table 3.9 were assumed to follow a normal distribution. The LSFs are established as:

$$g_1(X) = A \cdot \sigma - T = 0 \tag{3.23}$$

$$g_2(X) = f \cdot I_z / (h/2) \cdot \sigma - M = 0 \tag{3.24}$$

where A and σ are the cross-sectional area and yield stress in Tables (3.7, 3.8), respectively; T is the axial force, $f = 1.167$ is the ratio of the fully plastic moment to the yield moment, I_z is the moment of inertia of girder in Table 3.2, $h = 0.3$ m is the high point of the cross-section, and M is the bending moment of girders.

It is assumed that components (1–5) will fail due to tension, and components (6–7) will fail due to bending moment. For system reliability analysis, the system failure is defined using two different failure criteria. Criterion A: failure of any of components (1–7); Criterion B: failure of any two of the components (4–7).

In order to demonstrate the updating process in approximating the LSFs, the approximate curve and updated curve of girder component 6 are presented in Figure 3.11.

The counting procedures of Criterion A and B are shown in Figure 3.12, respectively. The compared result is shown in Table 3.10.

Table 3.10 presents the system reliability calculated by FELSYS-FEAP approach (Frangopol and Imai 2000) and the ASVR approach. The computational errors denote that the proposed ASVR approach possesses a reasonable accuracy. In this example, SVR is used to approximate LSFs in reliability analysis in order to reduce the number of runs in computation using a numerical method, such as FEM. Compared with the calculation in FEM, the SVR modeling and genetic algorithm-based optimization cost a slightly shorter time.

Overall, this example demonstrates that the proposed ASVR approach is an efficient method with reasonable accuracy for problems where closed-form failure functions are not available and the failure sequences exist in the structural system.

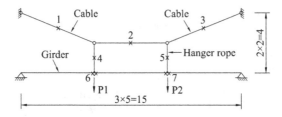

FIGURE 3.10　A suspended structure.

TABLE 3.8
Determined Parameters of the Suspended Structure

Type/unit	Component	Value
Cross-sectional area/mm²	Cable	225
	Hanger rope	100
	Girder	2450
Modulus/MPa	Cable	1.5×10^5
	Hanger rope	1.5×10^5
	Girder	2.0×10^5
Moment of inertia/mm⁴	Girder	3.19×10^7

TABLE 3.9
Random Variables of the Suspended Structure

Type/unit	Random variable	Mean value	Standard deviation
Load/kN	P_1	50	15
	P_2	50	15
Yield stress/MPa	σ_{cable}	1500	75
	σ_{hanger}	1000	50
	σ_{girder}	250	12.5

TABLE 3.10
Estimated System Reliability Utilizing the Two Approaches

	Probability of failure		
Criterion	RELSYS-FEAP	ASVR	Error(%)
A	0.02018	0.01969	−2.43
B	1.0514×10^{-7}	0.94420×10^{-7}	−10.20

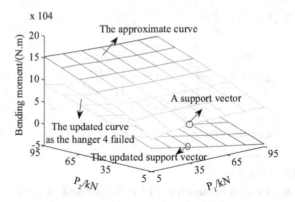

FIGURE 3.11   Updated curves and support vectors of the approximate LSF of girder 6.

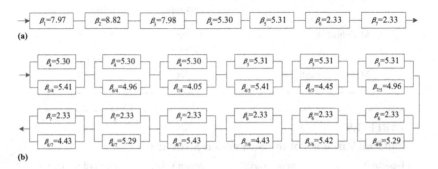

(a)

(b)

FIGURE 3.12   System reliability of Example 3.3. (a) Criterion A; (b) Criterion B.

3.4   VERIFICATION EXAMPLES

In order to demonstrate the effectiveness and feasibility of the proposed ASVR approach in structural reliability and system reliability estimation, two numerical examples were considered in this section.

3.5   APPLICATION EXAMPLE: A CABLE-STAYED BRIDGE

Due to the randomness of resistance and external loads, multiple failure modes and failure sequences exist in cable-stayed bridges. System reliability of cable-stayed bridges with uncertainties is challenging since the system analysis needs to include high order nonlinear implicit performance functions and various failure sequences.

3.5.1   PROJECT PROFILE AND THE FINITE ELEMENT MODEL

A pre-stressed concrete cable-stayed bridge on Luyu highway, China, is selected herein as a numerical analysis example. The bridge span arrangement is (210 + 420 + 410) m, as shown in Figure 3.13. The material of girders and towers are concrete, and the cables are steel strands. There are four traffic lanes. The system reliability of

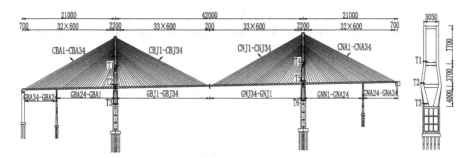

FIGURE 3.13 General arrangement diagram and series number of a cable-stayed bridge (unit: cm).

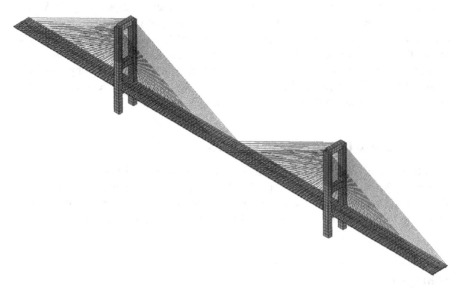

FIGURE 3.14 Finite element model of a cable-stayed bridge.

the bridge was estimated under vehicle loads during the operational period. The serial numbers of key elements are shown in Figure 3.13. The serial numbers of girders are GBA1~GBA34, GBJ1~GBJ34, GNJ1~GNJ34, GNA1~GNA34, those of the cable elements are CBA1~CBA34, CBJ1~CBJ34, CNJ1~CNJ34, CNA1 ~ CNA34, and those of the tower elements are T1~T6.

A 3D linear elastic finite element model of the bridge structure as shown in Figure 3.14 was constructed for reliability analysis. The bridge was modeled using the commercial finite element modeling code ANSYS. In the finite element model, the girders and towers were modeled by 470 Beam44 elements, and the cables were modeled by 272 Link 10 elements.

The main failure modes of cable-stayed bridges are bending failure of girders and towers, strength failure of cables, and large displacement failure of girders. These performance functions are implicit functions with multi-variables and high order nonlinearity. Thus, explicit performance functions should be deduced. Large axial forces exist in towers and girders of cable-stayed bridges and have an effect on the ultimate bearing capacity directly. Thus, girders and towers should be considered as axial-bending compounds. Their ultimate LSFs are

$$Z_i = 1 - \frac{P^i(X)}{P_u^i} - \frac{M^i(X)}{M_u^i} \tag{3.25}$$

$$Z_j = T_u^{\ j} - T^j(X) \tag{3.26}$$

$$Z_u = u_{max} - u(X) \tag{3.27}$$

where, X is the random variable vector. $P_u^{\ i}$ and $M_u^{\ i}$ are resistance of axial force and bending moment of girders and towers of the ith elements, respectively. $P^i(X)$ and $M^i(X)$ are response of axial force and bending moment of girders and towers at the ith elements under random loads, respectively. $T_u^{\ j}$ and $T^j(X)$ are strength and response of the jth elements of cables. u_{max} equal to $L/500$, according to *Design specifications of highway cable-stayed bridge JTJ027–96* in China, where L is the length of mid-span. $P^i(X)$, $M^i(X)$, $T^j(X)$ and $u(X)$ are implicit performance functions with high order nonlinearity and will be deduced by the SVR approach.

3.5.2 Random Parameters and Failure Modes

Vehicle loads of the cable-stayed bridges were considered in the operational period. Cheng and Xiao (2006) have observed that the cable-stayed bridge has a lower reliability if the vehicle loads are uniformly distributed in the mid-span of the bridge. Thus, the vehicle loads were simplified to be uniform loads in the mid-span. The

---

TABLE 3.11
Statistical Parameters of Random Variable for a Cable-Stayed Bridge

Type/unit	Position	Symbol	Distribution type	Mean	Standard deviation
Young modulus	Girder	E_1	Normal	3.64×10^4	3.64×10^3
/MPa	Tower	E_2		3.52×10^4	3.52×10^3
	Cable	E_3		1.95×10^5	1.95×10^4
Cross-section area/m²	Girder	A_1	Lognormal	20.846	1.042
	Tower	A_3		26.868	1.343
	Cable	A_5		1.4×10^{-4}	7.0×10^{-6}
Dead load/(kN.m⁻³)	Girder	γ_1	Normal	26.56	1.33
	Tower	γ_2		26.24	1.31
	Cable	γ_3		78.5	3.93
Moment of inertia/m⁴	Girder	I_1	Lognormal	18.598	0.930
	Tower	I_3		118.412	5.921
Dead load of deck pavement/(kN.m)	Bridge deck	q_1	Normal	132	6.6
Moving load/(kN.m⁻¹)	Bridge deck	q_2	Extreme value type I	63.5	6.35

yield strength of steel strand cables σ_{allow} = 1860 MPa, and the limit displacement u_{max} = 0.84 m.

There are many probabilistic variables existing in the cable-stayed bridge. The most used variables in bridge engineering are young modulus, cross-section area, moment of inertia, and the various loads. The statistical parameters of random variables of the bridge are shown in Table 3.11.

According to the mechanical characteristics of cable-stayed bridges, the structural system fails when the bending moment of tower and girder elements fail. Once the stress intensities of cables reach failure, we delete the failed elements and enter the next reliability analysis phase. In serviceability limit states, if the structural displacements reach the limit values, the structural system is deemed to have failed. In addition, when establishing the failure trees of cable-stayed bridges, different elements such as cables, girders, and towers, should be classified and the most possible failure elements are selected as the representative failure modes.

3.5.3 RESULTS AND DISCUSSION

Many potential failure modes exist in bridges. Regarding the cable-stayed bridge, the critical failure modes is exceedance of a predefined limit state such as displacement or yielding (Lee et al. 2014), but in the system reliability analysis the failure mode is defined as a sequence of failure components. These failure modes can warrant use of structural system in the static field. Based on the results discussed in Example 3.2, the system failure is defined using one failure criteria: failure of three components where a girder or a tower section is contained. According to the proposed ASVR approach, this analysis can be divided into two stages: component reliability and system reliability.

Yan and Chang (2009) demonstrated that stay cables are the most vulnerable components. Thus, the first stage is the reliability analysis of the cables. The calculated reliability indices of cables are shown in Figure 3.15. It can be noted that the reliability indices of cables are basically symmetrical. Cables in close proximity to the

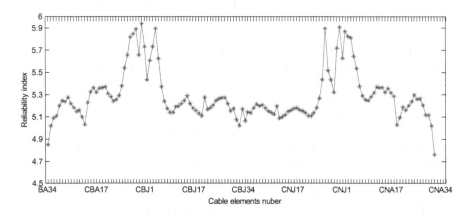

FIGURE 3.15 Reliability indices of cables.

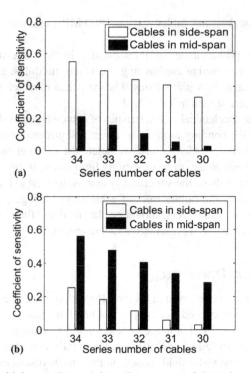

FIGURE 3.16   Sensitivity coefficients of bending moments for: (a) the pylon; (b) the girder.

towers are larger than other cables. The longer cables possess lower reliability indices, the minimum of which is CNA34 and the reliability index $\beta_{CNA34} = 4.76$.

The second stage is the analysis of system reliability of the cable-stayed bridge. A sensitivity analysis of components is considered to simplify establishing the failure sequences since components with the same type may be relevant. The sensitivity analysis of components is considered to be the potential failure compounds by deleting the failure components and recording the variations of towers in section T2 and girders in section GBJ1, GBJ34, and GNJ1. The coefficients of sensitivity are shown in Figure 3.16 (a, b).

Figure 3.16 indicates that cables in the side-span have a high influence on the bending moments of towers, and cables in mid-span have a high influence on the bending moments of girders in the mid-span. Furthermore, influence coefficients decrease with the decrease of cable numbers.

A sensitivity analysis of different components may provide the main failure sequences of the entire structure. Subsequently, the ASVR approach was applied to solve system reliability problem. The failure sequences were obtained and shown in Figure 3.17. The event tree and calculation chart are shown in Figure 3.18 (a, b).

As shown in Figure 3.18 (a, b), two main types of failure sequences can be concluded for a cable-stayed bridge. The first is strength failure of cables in the side-span (e.g., E_{CBA34}, E_{CNA34}, E_{CBA33}, E_{CNA33}) followed by bending moment failure of towers (e.g., T_2). The second is strength failure of cables in mid-span (e.g., E_{CBJ34}, E_{CNJ34},

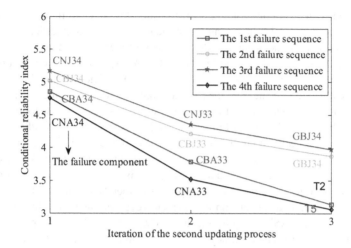

FIGURE 3.17 Failure sequences of the cable-stayed bridge.

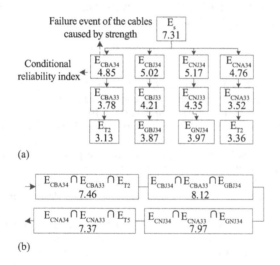

FIGURE 3.18 Event tree for hypothetic three-level system: (a) failure sequences; (b) series–parallel system.

E_{CBJ33}, E_{CNJ33}) followed by bending moment failure of girders (e.g., E_{CBJ34}, E_{CNJ34},). Furthermore, the reliability of cables greatly contributes to the system reliability.

3.6 CONCLUSIONS

A new ASVR method was proposed in this paper for structural system reliability assessment. Compared with traditional support vector regression (SVR), the proposed ASVR method involved updating two procedures of support vectors to approximate structural LSF. With the two updated processes, the traditional SVR has been improved and can be applied for engineering structural system reliability assessment. Two verification examples involving a nonlinear LSF and a geometrically nonlinear elastic truss illustrate the accuracy and efficiency of the proposed method. Finally, a pre-stressed concrete cable-stayed bridge with a mid-span of 420 m was considered to demonstrate the applicability of the proposed method. The numerical results of the cable-stayed bridge indicate that: the foremost failure sequence is the strength failure of side-span cables followed by the bending failure of towers; the secondary failure sequence is the strength failure of mid-span cables followed by the bending failure of mid-span girders. Cable reliability thus greatly contributes to the system reliability.

The ASVR method attempted to develop the traditional SVR method for structural reliability assessment. The application of this method is not only limited to cable-stayed bridges but can also be applied to other types of structures, such as continuous bridges and suspension bridges. However, it was found that many problems existing in this ASVR require further study, such as robustness in a large-scale complicated structure. The accuracy of the results in engineering structure is dependent on accuracy of finite element analysis. On the other hand, the multiple failure modes exist in cable-stayed bridges, such as stable, fatigue, dynamic, and so on. Therefore, it is recommended that further work be undertaken to investigate the ASVR method and its applications.

REFERENCES

Baldomir, A., Kusano, I. and Hernandez, S. et al. 2013. A reliability study for the Messina Bridge with respect to flutter phenomena considering uncertainties in experimental and numerical data. *Computers & Structures* 128: 91–100.

Botev, Z. I., L'Ecuyer, P. and Tuffin, B. 2013. Markov chain importance sampling with applications to rare event probability estimation. *Statistics and Computing* 23(2): 271–285.

Bucher, C. and Bourgund, U. 1990. A fast and efficient response surface approach for structural reliability problems. *Structural Safety* 7(1): 57–66.

Cardoso, J. B., Almeida, J. R. and Dias, J. M. et al. 2008. Structural reliability analysis using Monte Carlo simulation and neural networks. *Advances in Engineering Software* 39(6): 505–513.

Casciati, F., Cimellaro, G. and Domaneschi, M. 2008. Seismic reliability of a cable-stayed bridge retrofitted with hysteretic devices. *Computers & Structures* 86(17): 1769–1781.

Chan, C. L. and Low, B. K. 2012. Probabilistic analysis of laterally loaded piles using response surface and neural network approaches. *Computers and Geotechnics* 43: 101–110.

Cheng, J. and Li, Q. 2008. Reliability analysis of structures using artificial neural network based genetic algorithms. *Computer Methods in Applied Mechanics and Engineering* 197(45): 3742–3750.

Cheng, J. and Xiao, R. 2006. Application of inverse reliability method to estimation of cable safety factors of long span suspension bridges. *Structural Engineering and Mechanics* 23(2): 195–208.

Cheng, J., Zhang, J., Cai, C. and Xiao, R. 2007. A new approach for solving inverse reliability problems with implicit response functions. *Engineering Structures* 29(1):71–79.

Dai, H., Zhang, H. and Wang, W. 2012a. A support vector density-based importance sampling for reliability assessment. *Reliability Engineering & System Safety* 106: 86–93.

Dai, H., Zhang, H. and Wang, W. et al. 2012b. Structural reliability assessment by local approximation of limit state functions using adaptive Markov chain simulation and support vector regression. *Computer-Aided Civil and Infrastructure Engineering* 27(9): 676–686.

Degrauwe, D., Roeck, G. and Lombaert, G. 2009. Uncertainty quantification in the damage assessment of a cable-stayed bridge by means of fuzzy numbers. *Computers & Structures* 87(17): 1077–1084.

Deng, S. and Yeh, T. 2011. Using least squares support vector machines for the airframe structures manufacturing cost estimation. *International Journal of Production Economics* 131(2): 701–708.

Du, X. and Hu, Z. 2012. First order reliability method with truncated random variables. *Journal of Mechanical Design* 134(9): 091005. DOI:10.1115/1.4007150

Frangopol, D. and Imai, K. 2000. Geometrically nonlinear finite element reliability analysis of structural systems. II: Applications. *Computers & Structures* 77(6): 693–709.

Johnson, M. E. 2013. *Multivariate Statistical Simulation: A Guide to Selecting and Generating Continuous Multivariate Distributions.* John Wiley & Sons, New York, US.

Kang, S., Koh, H. and Choo, J. 2010. An efficient response surface method using moving least squares approximation for structural reliability analysis. *Probabilistic Engineering Mechanics* 25(4):365–371.

Kang, W., Song, J. and Gardoni, P. 2008. Matrix-based system reliability method and applications to bridge networks. *Reliability Engineering & System Safety* 93(11): 1584–1593.

Khatibinia, M., Fadaee, M. and Salajegheh, J. et al. 2013. Seismic reliability assessment of RC structures including soil–structure interaction using wavelet weighted least squares support vector machine. *Reliability Engineering & System Safety* 110: 22–33.

Kim, D., Ok, S. and Song, J. et al. 2013. System reliability analysis using dominant failure modes identified by selective searching technique. *Reliability Engineering & system safety* 119(11): 316–331.

Lee, Y. J., Lee, S. H. and Lee, H. S. 2014. Reliability assessment of tie-down cables for cable-stayed bridges subject to negative reactions: Case study. *Journal of Bridge Engineering* DOI: 10.1061/(ASCE)BE.1943-5592.0000717.

Lee, Y. and Song, J. 2012. Finite-element-based system reliability analysis of fatigue-induced sequential failures. *Reliability Engineering & System Safety* 108(12): 131–141.

Liang, H. S., Cheng, L. and Liu, S. 2011. Monte Carlo simulation based reliability evaluation of distribution system containing microgrids. *Power System Technology* 10: 76–81.

Lins, I., Moura, M. and Zio, E. et al. 2012. A particle swarm-optimized support vector machine for reliability prediction. *Quality and Reliability Engineering International* 28(2): 141–158.

Liu, Y., Lu, N. and Noori, M. et al. 2014. System reliability-based optimization for truss structures using genetic algorithm and neural network. *International Journal of Reliability and Safety* 8(1): 51–69.

Liu, Y., Lu N., Yin X., Noori M. 2016. An adaptive support vector regression method for structural system reliability assessment and its application to a cable-stayed bridge. *Proceedings of the Institution of Mechanical Engineers, Part O: Journal of Risk and Reliability* 230(2): 204–219.

Liu, L. and Zhu, S. 2013. Network flow and safety forecasting based on least squares SVR optimized by particle swarm optimization. *Journal of Computational Information Systems* 9(21): 8709–8716.

Moura, M., Zio, E. and Lins, I. et al. 2011. Failure and reliability prediction by support vector machines regression of time series data. *Reliability Engineering & System Safety* 96(11): 1527–1534.

Murotsu, Y. 1984. Automatic generation of stochastically dominant modes of structural failure in frame. *Structural Safety* 2(1): 17–25.

Noh, Y., Choi, K. and Du, L. 2009. Reliability-based design optimization of problems with correlated input variables using a Gaussian Copula. *Structural and Multidisciplinary Optimization* 38(1): 1–16.

Suykens, J. and Vandewalle, J. 1999. Least squares support vector machines classifiers. *Neural Network Letters* 19(3): 293–300.

Tan, X., Bi, W. and Hou, X. et al. 2011. Reliability analysis using radial basis function networks and support vector machines. *Computers and Geotechnics* 38(2): 178–186.

Tang, Q. and Zhang, C. 2013. Data Processing System (DPS) software with experimental design, statistical analysis and data mining developed for use in entomological research. *Insect Science* 20(2): 254–260.

Vanli, O. and Jung, S. 2014. Statistical updating of finite element model with Lamb wave sensing data for damage detection problems. *Mechanical Systems and Signal Processing* 42(1): 137–151.

Vapnik, V. 2000. *The Nature of Statistical Learning Theory*. 2 ed. New York: Springer-Verlag.

Wei, Z., Tao, T. and Zhuoshu, D. et al. 2013. A dynamic particle filter-support vector regression method for reliability prediction. *Reliability Engineering & System Safety* 119: 109–116.

Wu, C. H., Tzeng, G. H. and Lin, R. H. 2009. A novel hybrid genetic algorithm for kernel function and parameter optimization in support vector regression. *Expert Systems with Applications* 36(3): 4725–4735.

Yan, D. and Chang, C. 2009. Vulnerability assessment of cable-stayed bridges in probabilistic domain. *Journal of Bridge Engineering* 14(4): 270–278.

Zhang, W., Cai, C. and Pan, F. 2012. Fatigue reliability assessment for long-span bridges under combined dynamic loads from winds and vehicles. *Journal of Bridge Engineering* 18(8): 735–747

4 System Reliability Evaluation of In-service Cable-stayed Bridges Subjected to Cable Degradation

Naiwei Lu

Changsha University of Science and Technology, China

Yang Liu

Hunan University of Technology, China

Michael Beer

Leibniz Universität Hannover, Germany

CONTENTS

4.1 INTRODUCTION

The cables in a cable-stayed bridge connecting bridge decks and pylons are critical components ensuring the long-span capability of bridge girders. The cables provide high degrees of redundancy for the structure, thereby making the structural system stronger and more robust. However, cables are particularly vulnerable to fatigue damage and atmospheric corrosion during the service period (Yang et al. 2013; Lu and He 2016), which contributes to the risk of bridge failure. Stewart and Al-Harthy (2008) indicated that a spatial variability in corrosion of a steel bar led to a 200% higher failure probability. In practice, fatigue damage and corrosion are normal phenomena for steel strands or parallel wires in a stay cable. An inspection conducted by Mehrabi et al. (2010) showed that 39 of the 72 cables of the Hale Boggs Bridge were critically in need of repair or replacement after a 25-year service period. Although a cable-stayed bridge is designed with sufficient conservatism, the system is still vulnerable from cable degradation. Uncertainties in structural parameters also contribute to the system safety risk (Li et al. 2015). Therefore, the influence of cable degradation and associated uncertainties on the system safety of cable-stayed bridges deserves investigation.

The most evident form of cable degradation is the nonuniformly distributed reduction in the cross-sectional area of a cable due to environmental corrosion (Yang et al. 2012; Xu et al. 2016). Loss in the cable cross-sectional area directly decreases the ultimate strength of a cable in a series–parallel system (Xu and Chen 2013). In addition to the continuous reduction in cable strength, cable degradation also increases the risk of cable rupture. During a cable rupture, the dynamic force acting on the remaining system can lead to failure of the adjacent cables and girders (Wolff and Starossek 2009). This phenomenon has been summarized in terms of progressive collapse (Marjanishvili 2004) or zipper-type collapse (Starossek 2007). To avoid the propagation of cable loss in a cable-stayed bridge, existing cable-stayed bridges are mostly designed with high degrees of static indeterminacy and conservatism (Aoki et al. 2013). However, the effect of cable loss on bridge safety, e.g., vehicle–bridge–wind interaction (Zhou and Chen 2014; Zhou and Chen 2015) and dynamic amplification factors (Mozos and Aparicio 2011), still attract the attention of researchers.

Significant research has been devoted to component-level reliability evaluation of cable-stayed bridges, e.g., Cheng and Xiao (2005) and Li et al. (2012). However, a cable-stayed bridge is a complex system composed of multiple components, and the failure mode of the bridge is associated with numerous components. Taking into account structural uncertainties and cable degradation, the system safety evaluation of cable-stayed bridges is more complicated. Since each cable provides a degree of redundancy for a cable-stayed bridge, the bridge failure can be defined as several components connected in series or in parallel. Hence, system reliability theory is appropriate for this issue (Estes and Frangopol 2001). The most influential approaches are the β-unzipping method (Thoft-Christensen and Murotsu 1986), the branch-and-bound method (Lee and Song 2011), and the selective searching approach (Kim et al. 2013). Initially, Bruneau (1992) identified nine potential failure patterns of a short-span cable-stayed bridge via the global ultimate capacity approach. Li et al. (2010) utilized the β-unzipping method to evaluate the system reliability and found that the cables were the critical components impacting the system reliability of a cable-stayed bridge. Zhu and Wu (2011) used the Bayesian updating method to evaluate the system reliability of a cable-stayed bridge with inspection information. To make the computation more efficient, Liu et al. (2016) developed a machine learning approach for system reliability evaluation of complex structures and applied it to cable-stayed bridges. Jia et al. (2016) utilized a direct differential method in a stochastic finite element model to simulate the response gradient of the structural system of a cable-stayed bridge. However, to the best of the authors' knowledge, research on system reliability evaluations of cable-stayed bridges subject to cable degradation remains inadequate.

This chapter evaluated the system reliability of in-service cable-stayed bridges subjected to cable degradation. The strength reduction of parallel wire cables due to fatigue damage and corrosion are considered in a parallel–series system. An intelligent machine learning approach is presented for formulating the structural failure paths and evaluating the lifetime system reliability of cable-stayed bridges. Finally, two cable-stayed bridges, including an ancient short-span bridge and a modern long-span bridge, were selected as prototypes to investigate the influence of cable degradation on the structural failure modes and failure paths. Parametric studies associated with random variables and correlation coefficients were conducted.

4.2 MODELING STRENGTH REDUCTION OF PARALLEL WIRE CABLES

Both parallel wires and strands are conventional types of stay cables for a cable-supported bridge. A parallel wire cable consists of parallel, straight, round wires inside a polystyrene pipe. In addition to the material characteristics, the strength of parallel wires is also associated with the length and the number of wires (Nakamura and Suzumura 2012).

4.2.1 EFFECTS OF CABLE LENGTH AND NUMBER OF WIRES

First, this study conducted mathematical properties modeling of wire cables considering the effect of wire length and number of wires. In general, a parallel wire cable can be modeled in a series–parallel system, as shown in Figure 4.1, where λ is a scale describing the corrosion state of a cable, and L and L_0 are total length and correlation length of the wire, respectively. In Figure 4.1, each wire can be simulated as a series system, and the wires work together as a parallel system. In the series system, the individual wires of a stay cable can be simulated with correlative elements depending on the length of the wire. The material properties and defects in the wires are considered by a correlation length L_0 in the series system. The cable strength decreases in association with either shorter correlation lengths or longer wire lengths. Therefore, both the correlation and wire lengths have been considered in the parallel–series model. A distribution function of the strength of a wire can be written using a Weibull cumulative distribution function (Weibull 1949):

$$F_Z(z) = 1 - \exp\left[-\lambda\left(\frac{z}{u}\right)^k\right] \tag{4.1}$$

where z is the strength of a wire, u and k are unknown scale and shape parameters in the Weibull distribution function that can be estimated from ultimate capacity tests via

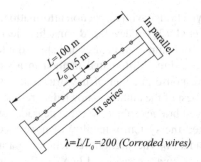

FIGURE 4.1   A parallel–series system of a stay cable.

the maximum likelihood method. An experimental study of a wire ($L = 100$ m) conducted by Faber et al. (2003) indicated that the mean values of the wire strength for
undamaged cable ($\lambda = 3$) and corroded wire ($\lambda = 200$) were 1748 MPa and 1650 MPa,
respectively. Faber concluded that the variability in the wire strength is negligible.

In addition to the series system of an individual wire, a stay cable consists of
numerous wires in parallel, as shown in Figure 4.1. In the parallel system, increasing
the number of wires (n) in a cable will reduce the mean strength of each wire, which
is the so-called Daniel's effect (Daniels 1945). If n is large enough, the strength of a
parallel wire follows a normal distribution with the mean value and standard deviation (Faber et al. 2003)

$$E(n) = nx_0\left(1 - F_Z(x_0)\right) + c_n \tag{4.2a}$$

$$D(n) = x_0\left[nF_Z(x_0)\left(1 - F_Z(x_0)\right)\right]^{1/2} \tag{4.2b}$$

where, $c_n = 0.966an^{1/3}$, $a^3 = \dfrac{f_Z^2(x_0)x_0^4}{(2f_Z(x_0) + x_0 f(x_0))}$, $x_0 = \left[l\dfrac{L_0}{Lk}\right]^{1/k}$ σ_u, σ_u is the mean
strength of a wire, and $f_Z(x_0)$ is a Weibull probability density function.

Using the data of the tensile strength tests of 30 wires conducted by Faber et al.
(2003) as an example, the mean value of the wire specimen strength $\sigma_u = 1788.7$
MPa, the Weibull parameter $k = 72.62$, and the scale factor is assumed to be $\lambda = 1, 3$,
and 10. The reduction curves for the mean strength of a cable system composed of 10
to 300 wires are shown in Figure 4.2.

It can be observed in Figure 4.2 that the wire strength decreases rapidly with the
initial increase in the number of wires, an increase in the number of wires from 10 to 300
leads to a 4.3% reduction in the wire strength. In addition, the wire strength decreases
with a larger scale factor or a smaller correction length will reduce the wire strength. The
deviation due to the Daniels effect is negligible, similar to the length effect.

4.2.2   Effect of Fatigue-Corrosion

Since high stress cables are prone to corrosion, stay cables become rusty and frac-
tured under conditions of long-term exposure to corrosion and cyclic stresses Li et al.

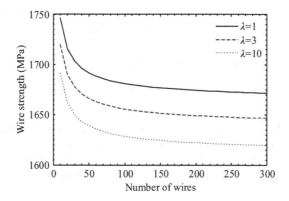

FIGURE 4.2 Wire strength in a stay cable impacted by the number of wires.

2014. Thus, cable corrosion is a common phenomenon in existing cable-supported bridges. The corrosion of cable wires takes different forms, including stress corrosion cracking, pitting, corrosion fatigue, and hydrogen embrittlement, which lead to reductions in the strength and ductility of wires and subsequently a reduced lifetime Mahmoud 2007. This study considers cable degradation resulting from atmospheric corrosion and fatigue damage.

In the concept of fatigue damage accumulation, it is assumed that the wire with the shortest failure time breaks first in the system. Subsequently, the stress is redistributed to the remaining wires. Therefore, the probability distribution function of the initial failure time is written as Maljaars and Vrouwenvelder 2014:

$$F_N\left(N, \sigma_{eq}\right) = 1 - \exp\left[-\left(\frac{\sigma_{eq}}{r_c}\right)^\alpha \left(\frac{N}{K}\right)^{\frac{\alpha}{m}}\right] \qquad (4.3)$$

where σ_{eq} and N are the equivalent stress range and the number of stress cycles, respectively, and can be estimated considering actual traffic loads; α, m and K are unknown coefficients that can be estimated by experimental tests; $r_c = \left(cnA_0\right)^{-\frac{1}{\alpha}}$ where n is the number of wires of the specimen, A_0 is the cross-sectional area of the wire, and c is an unknown parameter; $K = K_0\sqrt{1 - \dfrac{m_s}{m_z}}$ where K_0 can be estimated from experiments of wires with a known strength and stress, m_s and m_z are the mean stress and the mean strength, respectively. Note that since the cable strength degradation model is associated with loading scenarios, the distribution may not be identically and independently approximated.

Based on the assumption proposed by Faber et al. (2003), the parameters for the cable are $n = 200$, $\sigma_u = 1789$ MPa, $\lambda = 3$ and $\lambda = 200$ for non-corroded and corroded wires, respectively; the parameters for the loading are $N = 1.5 \times 10^6$ cycles per year, $\sigma_{eq} = 30$ MPa. In addition, the estimated parameters in Equation (4.2a and 4.2b) are

$m = 1.50$, $\alpha = 2.76$, and $K_0 = 1.19 \times 10^6$ MPa. Degradation functions of the cable strength for the uncorroded cables and corroded cable are

$$y_1(t) = -1.5 \times 10^{-5} t^2 - 3.2 \times 10^{-3} t + 0.998 \tag{4.4a}$$

$$y_2(t) = -4.7 \times 10^{-4} t^2 - 2.4 \times 10^{-3} t + 0.996 \tag{4.4b}$$

where t is time of year. The uncorroded cable is associated with only fatigue damage, and the corroded cable is associated with both fatigue and corrosion. It is observed that the strength coefficients of the stay cable associated with fatigue and fatigue-corrosion effects during the 20-year service period are 0.928 and 0.751, respectively.

4.2.3 PROBABILISTIC MODELING OF CABLE STRENGTH

As mentioned before, Faber et al. (2003) presented the procedures for probability modeling for cable strength using ultrasonic inspections and experimental results. However, these results are hypothetical, since the experimental data were mostly assumed. This study considers the actual strength of the specimens of 69 corroded wires and 13 uncorroded wires in the stay cables of a highway bridge in China (Li et al. 2012). The wire specimens of the cables in service for around 20 years should have a free length of 500 mm (Lu et al. 2018). The evaluated probability models of the wire specimens are shown in Figure 4.3.

In the present study, consider a stay cable with a length of 232 m that will be adopted in the following case study. The cable consists of $n = 243$ parallel steel wires, and each wire has a diameter $\varphi = 7$ mm. The design strength of the cable is 1766 MPa. Initially, the probability model of the strength of an individual wire was evaluated based on Equation (4.2a and 4.2b) and Figure 4.3. The results are shown in Figure 4.4.

As observed from Figure 4.4, the mean value of the cable strength had decreased by 32% and 13%, respectively. However, the standard deviation had decreased less than 2%. The negligible deviation in cable strength due to the cable length effect and the Daniels effect is in agreement with Faber et al. (2003). Based on the probability

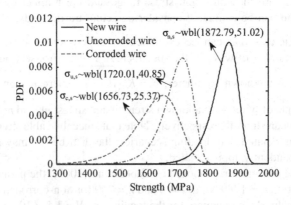

FIGURE 4.3 Probability density of the strength of the specimens.

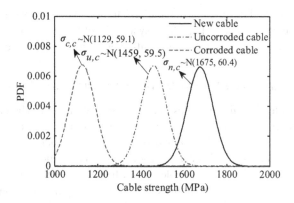

FIGURE 4.4 Probability distribution of a prototype cable.

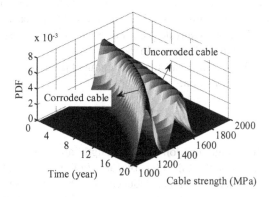

FIGURE 4.5 Time-variant probability densities of the cable strength.

model of the prototype cable as shown in Figure 4.4, the time-variant probability model was evaluated considering the degradation function shown in Equation (4.4a and 4.4b). The time-variant degradation model of the cable in a lifetime is shown in Figure 4.5.

The probabilistic degradation model in Figure 4.5 was evaluated based on the following assumptions: (1) the probabilistic model of new, uncorroded, and corroded wires refers to the test data of 82 wires conducted by Li et al. (2012); (2) the relationship between the probability models of wires and cables were considered in a parallel–series system; (3) the degradation tendency of the cable in the 20-year service period refers to Faber et al. (2003).

4.3 A COMPUTATIONAL FRAMEWORK FOR SYSTEM RELIABILITY EVALUATION OF CABLE-STAYED BRIDGES

Compared with girder-type bridges, a cable-stayed bridge exhibits unique characteristics, such the cable sag and beam-column effects and nonlinear behaviors Freire et al. 2006. A cable-stayed bridge is a complex system consisting of multiple components connected in series or in parallel. These components work together as a system to ensure bridge safety. Therefore, system-level behaviors should also be further considered in addition to the component-level behaviors illustrated above.

4.3.1 COMPONENT-LEVEL FAILURE MODE OF A CABLE-STAYED BRIDGE

In general, the potential failure modes of a cable-stayed bridge are bending failure of towers and girders, strength failure of cables, and stability failure of pylons. This study focused on the bending failure and strength failure modes, while the stability failure of pylons (Nariman 2017) can be readily amended in the systems model as an additional component. Long-span cable-stayed bridges exhibit nonlinear behaviors under heavy traffic loading. The nonlinearities, including the cable sag and beam-column effects, and large displacement, should be considered when formulating the limit state functions.

For the strength failure of a stay cable, the long cable sag effect resulting from self-weight is an evident nonlinearity. The conventional approach for considering this nonlinearity is the Ernst equivalent elastic modulus. Large displacement of the girders also leads to nonlinear behavior in the stay cable. These nonlinearities can be considered in a time-history analysis in a finite element program and can be captured in a nonlinear limit state function:

$$Z_{Cable}^i = T_u^i\left(t\right) - T^i\left(X\right) \tag{4.5}$$

where Z_{Cable}^i is the limit state function for the strength failure of the ith cable, $T_u^i\left(t\right)$ is the time-variant wire strength of the ith cable, and $T^i(X)$ is the actual stress of the jth cable. The cable resistance term can be deduced from the models of the series–parallel system and the fatigue and corrosion effects. The cable force term can be approximated by learning machines with the samples simulated by finite element analysis.

Another significant failure mode for cable-stayed bridges is the bending failure of girders and pylons. Due to the high stress in stay cables, the beam-column effect is a unique factor impacting the mechanical behavior of the beams and pylons, especially in their bending failure state. In fact, the relationship between bending moment and axial force will affect the component stiffness coefficient and the internal forces. The beam-column interaction is a second-order effect and can be conveniently considered by utilizing stability functions (Yoo et al. 2010). Assuming a hollow rectangular section, where the neutral axis in the ultimate state within the webs, the axial bending interaction curve can be evaluated via a sample plastic analysis and is written as follows (Yoo et al. 2012):

$$\frac{M}{M_P} = 1 - \left(\frac{P}{P_P}\right)^2 \frac{A^2}{4wZ_x} \tag{4.6}$$

where M is the applied moment, M_p is the plastic moment capacity in the absence of axial loads, P is the applied axial force, P_p is the plastic axial force capacity in the absence of applied moment, w is the web thicknesses, and Z_x is the bending plastic modulus. The beam-column effect can also be captured by a learning machine, which will be described later in detail.

4.3.2   SYSTEM-LEVEL FAILURE SEQUENCES AND SUBSYSTEM UPDATING

Since a cable-stayed bridge is a statically indeterminate system composed of girders, cables, and pylons, the failure of a component will result in the redistribution of the stress and strain within the system. If the structural subsystem still has the capacity to withstand the load, more and more components will fail with increases in the load. Finally, the structure becomes an unstable system, and the structure will collapse or fail to meet a requirement. In the above procedures, the system reliability can be modeled by combining the failed components into a parallel–series system. In this area, the conventional approaches of identifying the failure sequences are associated with the selecting and searching technologies. One of the technologies used to search for potential failure components is the branch-and-bound method known as the β-unzipping method (Liu et al. 2014).

The β-unzipping method uses the reliability index of each component to search for potential failure components. The removing of each potential failure component branches the initial structure into different substructures. The connection of the identified failed components into a parallel system forms a failure sequence that is similar to a zipper. The searching procedure, therefore, is actually an unzipping process. Suppose a system composed of n components, where k-1 components (denoted as r_1, $r_2,\dots, r_{k-1}$) failed. The conditional reliability index of the kth component is written as follows (Thoft-Christensen and Murotsu 1986):

$$\beta_{r_k/}^{(k)} = \beta_{r_k/r_1,r_2,\cdots r_{k-1}}^{(k)} = \Phi^{-1}\left[P\left(E_{r_k/}^{(k)}\right)\right] \tag{4.7}$$

where $E_{r_k/}^{(k)}$ is the event of the kth component failure at the kth failure phase, $P()$ denotes the probability of the event, $\Phi^{-1}()$ is an inverse cumulative distribution function, and $\beta_{r_k/r_1,r_2,\cdots r_{k-1}}^{(k)}$ is a conditional reliability index of the kth potential component at the kth failure phase. The condition to be the potential failure component is as follows (Liu et al. 2016):

$$\beta_{r_k/}^{(k)} \le \beta_{\min}^{(k)} + \Delta\beta \tag{4.8}$$

where $\beta_{\min}^{(k)}$ is the minimum reliability index at the kth failure phase, and $\Delta\beta$ is the increasing interval of the reliability index. Thoft-Christensen and Murotsu (1986) suggested that $\Delta\beta$ can be treated as 3 during the first searching process and 1 for the latter processes, as an efficient searching and selecting scheme.

The most critical procedure in the β-unzipping method is to break the initial system into subsystems that is called system updating. System updating is associated with the failure mode of the potential component. In structural engineering, the conventional approach for system updating involves adding a plastic hinge at the failure position to account for bending failure. However, because a cable-stayed bridge has multiple different failure modes as illustrated above, special attention should be paid to the system updating. The following assumptions for system updating for cable-stay bridges can be referred from the previous studies Bruneau 1992. First, for a brittleness failure mode, such as a cable rupture, the subsystem can be updated by directly removing the failed component. Second, for a ductile failure mode, such as the bending failure of girders and pylons, the subsystem can be updated by adding a plastic hinge at the failure location. The system failure of a cable-stayed bridge can be determined by the load-carrying capability of the entire structure or any other criterion, such as the serviceability.

In summary, the β-unzipping process is described as follows. First, conduct a component-level reliability analysis for each structural element, then choose the minimum reliability index $\beta_{min}^{(i)}$. Second, select the potential failure components with the reliability index in the interval $[\beta_{min}^{(i)}, \beta_{min}^{(i)} + \Delta\beta]$. Third, branch the initial system into different subsystems according to the system updating criterion by separately removing the identified failed components. Next, repeat the above procedures until the structure collapses or fails to meet a specified requirement. Finally, combine these reliability indices of identified components into a parallel–series system that is similar to an event tree, where the system reliability can eventually be evaluated.

As observed from the above deduction, the beam-column effect is considered in Equation (4.3), and the failure sequence searching criteria is shown in Equations (4.4a and 4.4b) and (4.5). In addition to these unique characteristics illustrated above, the cable degradation will add to the time-consuming computation of the system reliability. Therefore, special attention should be paid to utilize an efficient computational framework.

4.3.3 STRUCTURAL SYSTEM RELIABILITY EVALUATION VIA MACHINE LEARNING

Due to the characteristics and cable degradation of the cable-stayed bridge, an efficient framework should meet the following requirements. Initially, the computing accuracy is an essential requirement, because the components have nonlinear mechanical behavior and extremely low failure probability. On this issue, the conventional first-order-second-moment (FOSM) approach cannot capture the structural nonlinear behavior, and the conventional second-order response surface method (RSM) is incapable of capturing the structural higher-order nonlinear behavior (Alibrandi et al. 2015; Liu et al. 2016). Subsequently, since searching for the dominant failure sequence is a time-consuming process, computational efficiency is another essential requirement. Based on the above formulations, this study utilized a machine learning approach based on the adaptive support vector regression (ASVR) approach proposed by Liu et al. (2016). The framework of the intelligent algorithm was improved for special application by taking into account the cable degradation. The flowchart of the framework is shown in Figure 4.6.

As depicted in Figure 4.6, the main procedures in the flowchart consist of two aspects including the system reliability evaluation based on the ASVR approach and the updating of the cable degradation model. Details of the applications of the support vector regression (SVR) and ASVR approaches in structural system reliability evaluation can be found in Dai et al. (2012) and Liu et al. (2016). This article mainly

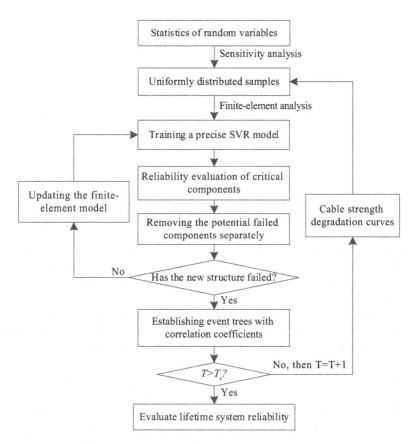

FIGURE 4.6 Flowchart of the proposed ASVR approach.

describes the procedures associated with cable degradation. At first, the cable strengths were taken as initial values to carry out the system reliability evaluation step by step. Subsequently, the cable strength model was updated, and the component and system reliability then reevaluated. It is worth noting that the event tree should be rebuilt when the cable strength changes, because cable degradation may change the potential failed components comprising the failure sequence. Finally, if the service period, T_s, is reached, the entire procedure will stop and output the lifetime system reliability indices. The cable degradation-induced decrease in the system reliability will be reflected by the time-variant system reliability indices.

The critical step (as depicted in Figure 4.6) is resampling the training samples after the cable strength is updated. This procedure suggests that it is inappropriate to only update the reliability index of individual cables without the reevaluation of the failure sequences. Instead, the seemingly additional computational effort is essential to capture the main failure sequences. Such deduction will be demonstrated in the case study.

As mentioned above, the introduction of the cable degradation leads to additional computational efforts. To make the computation more efficient, a graphical user interface (GUI)-based program entitled "Complex Structural Reliability Analysis

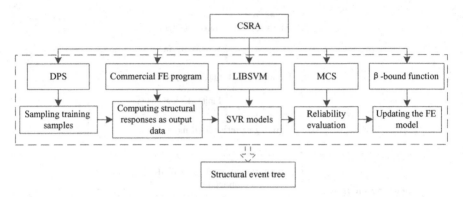

FIGURE 4.7   Flowchart of the complex structural reliability analysis (CSRA) program.

software V 1.0" (CSRA 2013) was developed based on the framework via MATLAB (2010). This general program is based on two commercial programs, namely, MATLAB and ANSYS. The main procedures of the CSRA program are summarized in Figure 4.7.

In Figure 4.7, the Data Processing System (DPS) Tang and Zhang 2013 is utilized to generate uniformly distributed samples that will be used for training SVR models. The LIBSVM (Library for Support Vector Machines) (Chang and Lin 2011) is a MATLAB program package. The MCS (Monte Carlo Simulation) can be a direct or importance sampling. Compared with the direct MCS, importance sampling uses a weighting function to concentrate the sampling around the most probable point (MMP) rather than around the mean vector (Dai et al. 2012). In this manner a large number of sample points fall into the failure domain making the sampling much more efficient. The β-bound function refers to Equations (4.6) and (4.7). At the end of the procedure, the finite element model is updated, and the component reliability is then reevaluated step by step. Eventually, the system reliability can be evaluated in a series–parallel system.

4.4   CASE STUDY OF A SHORT-SPAN CABLE-STAYED BRIDGE

An ancient short-span cable-stayed bridge shown in Figure 4.8 is presented to investigate the influence of cable degradation on the bridge system reliability. The cable-stayed bridge has a single pylon and two stay cables on each side. The distance between the cable anchors in the girders is 30 m. More details regarding the material and section properties and performance functions can be found in Bruneau (1992). In this case study, the structural mechanical behavior was assumed to be linear and elastic in accordance with the results provided by Bruneau (1992).

In general, cables are considered brittle because the rupture of a stay cable is transient. The concrete girders and towers in long-span bridges are considered ductile because the prestressed structures are able to have large deflections. The structural system failure was defined by a plastic collapse mechanism, which was identified by the plastic hinge locations and plastic capacities. The potential failure locations are shown in Figure 4.6. The points G1 ~ G11 of the girders and the points of T_1 and T_2

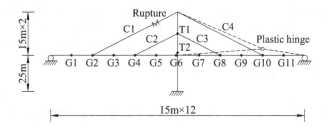

FIGURE 4.8 Dimensions and failure modes of a short-span cable-stayed bridge.

of the pylon are critical to bending failure, and the points C1 ~ C4 of the stay cables are critical to sudden rupture.

From a system-level point of view, if a cable is ruptured, the substructure can be updated by removing the ruptured cables directly. If bending failure of a girder or a pylon occurs, a plastic hinge can be added to the location where the bending failure occurs. It is acknowledged that the structural stiffness and resistance can change at any step, which means that the remaining structural elements will form a new structural system. At the end of the processes, the progression along the failure sequences will stop when the failure probability of the final component is expected to be extremely high. A two-level event tree was adopted by Bruneau (1992) to describe the particle event for the cable-stayed bridge. In the present study, the cable strength coefficients shown in Figure 4.2 were used to update to the limit state functions provided by Bruneau (1992). Based on the above assumptions, the event trees of the bridge system were evaluated considering no degradation and a degradation coefficient of 20%, respectively. Figure 4.9 plots the two-level failure sequences of the system.

The following conclusions can be deduced from Figure 4.9. First, the dominant failure sequence of the initial bridge is the Hinge @ G10 followed by the Hinge @ G2. However, the cable degradation shifts of the dominant failure to the Cable @ C2 followed by the Hinge @ G6. Secondly, a 20% reduction in the cable strength results in a rapid increase in the failure probability of the C2 cable from 0.154×10^{-7} to 0.243×10^{-4}. Finally, the decrease of the cable reliability results in the structural system reliability decrease from 4.67 to 3.92. In conclusion, the cable degradation not only decreases the reliability of a cable but also has a significant impact on the structural dominant failure mode and the system reliability.

To investigate the lifetime degradation of the cable strength on the system reliability of the bridge, the system reliability was reevaluated based on the proposed framework. Figure 4.10 plots the results of the time-variant system reliability of the bridge over a 20-year service period.

Figure 4.10 shows that the system reliability indices have a similar tendency to the cable strength models. The reliability index accounting for fatigue and corrosion decreases rapidly compared to that caused by fatigue. However, the fatigue-corrosion effect leads to a sudden decrease in the reliability index starting in the thirteenth year. This phenomenon can be explained by the event tree shown in Figure 4.9, where the cable failure becomes the dominant failure mode as the cable strength decreases to a critical value. The failure probability of a cable due to fatigue-corrosion effect will be larger than that of the hinge at the G10 point from the thirteenth year of service period. It can be derived that the continuous cable degradation has resulted in a transformation of the domain failure component from the girder to the cable, which leads to a sudden decrease in the system reliability index. However, this phenomenon is not observed in association with the fatigue-induced reliability decrease, because the cable strength has not decreased to the limit.

4.5 CASE STUDY OF A LONG-SPAN CABLE-STAYED BRIDGE

4.5.1 PROTOTYPE BRIDGE

A long-span cable-stayed bridge (Liu et al. 2016) crossing the Yangtze River in Sichuan province, China, is selected herein as a prototype to investigate the influence of cable degradation on the system reliability. The bridge has two pylons with

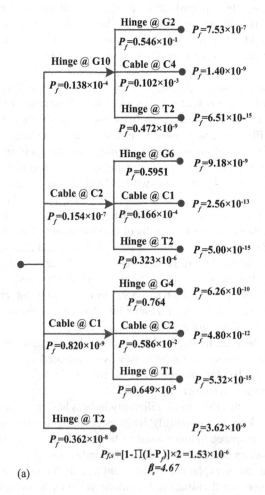

(a)

FIGURE 4.9 Two-level event trees of the short-span cable-stayed bridge: (a) without cable degradation.

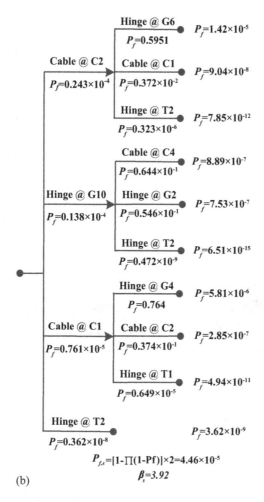

Hinge @ G6
$P_f=0.5951$
$P_f=1.42\times10^{-5}$

Cable @ C2 | **Cable @ C1**
$P_f=0.243\times10^{-4}$ | $P_f=0.372\times10^{-2}$
$P_f=9.04\times10^{-8}$

Hinge @ T2
$P_f=0.323\times10^{-6}$
$P_f=7.85\times10^{-12}$

Cable @ C4
$P_f=0.644\times10^{-1}$
$P_f=8.89\times10^{-7}$

Hinge @ G10 | **Hinge @ G2**
$P_f=0.138\times10^{-4}$ | $P_f=0.546\times10^{-1}$
$P_f=7.53\times10^{-7}$

Hinge @ T2
$P_f=0.472\times10^{-9}$
$P_f=6.51\times10^{-15}$

Hinge @ G4
$P_f=0.764$
$P_f=5.81\times10^{-6}$

Cable @ C1 | **Cable @ C2**
$P_f=0.761\times10^{-5}$ | $P_f=0.374\times10^{-1}$
$P_f=2.85\times10^{-7}$

Hinge @ T1
$P_f=0.649\times10^{-5}$
$P_f=4.94\times10^{-11}$

Hinge @ T2
$P_f=0.362\times10^{-8}$
$P_f=3.62\times10^{-9}$

$$P_{f,s}=[1-\prod(1-Pf)]\times2=4.46\times10^{-5}$$
$$\beta_s=3.92$$

(b)

FIGURE 4.9 **(Continued)** Two-level event trees of the short-span cable-stayed bridge: (b) with a cable degradation coefficient of 20%.

double-sided cables forming a fan pattern. The pylon and segmental girders are connected by 34 pairs of cables on two sides. The dimensions of the bridge are shown in Figure 4.11, where C_s^i and C_m^i denote the ith pair of cables in the side- and mid-spans, respectively; G_s^j and G_s^j denotes the jth pair of girders in the side- and mid-span sides, respectively; and P_1, P_2, and P_3 denote the critical bending failure points at the pylon and the girder.

The diameter of a single parallel steel wire is $\varphi = 7$ mm, and the physical properties of the wire were taken from the specifications of MOCAT (2010). The modulus of elasticity of the cables was estimated via Ernst's equation. The physical properties of the four longest cables are shown in Table 4.1, where N_s is the number of wires in a stay cable, A_s is the cross-sectional area of the cable, $E_{s,Ernst}$ is the equivalent elastic modulus of the cable, and T_0 is the initial cable force. Design values of the

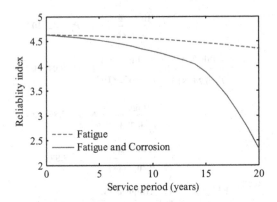

FIGURE 4.10 System reliability of the short-span bridge subjected to cable degradation.

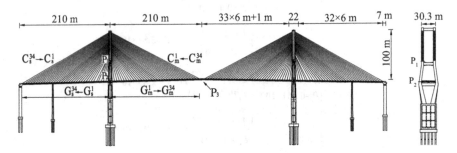

FIGURE 4.11 Dimensions of the cable-stayed bridge: (a) elevation view; (b) side view.

TABLE 4.1
Physical Properties of the Four Longest Stay Cables

Cables	N_s	A_s (m²)	$E_{s,Ernst}$ (MPa)	T_0 (kN)
C_s^{34}, C_m^{34}	253	9.73×10^{-3}	1.864×10^5	6440
C_s^{33}, C_m^{33}	241	9.28×10^{-3}	1.860×10^5	6066
C_s^{32}, C_m^{32}	241	9.28×10^{-3}	1.852×10^5	5665
C_s^{31}, C_m^{31}	223	8.58×10^{-3}	1.841×10^5	5580

parameters of the long-span cable-stayed bridge are shown in Table 4.2. The four-lane vehicle load was simplified as a uniform load on the mid-span girders following a normal distribution with the mean value of 63.5 kN/m and a 6.35 standard deviation. The purpose of the case study is to demonstrate the effectiveness of the proposed computational framework and to investigate the influence of cable degradation on the system reliability of long-span cable-stayed bridges.

TABLE 4.2
Design Values of the Parameters of the Long-Span Cable-Stayed Bridge

Variable	Mean value	Description
E_c	3.45×10^4 MPa	Modulus of elasticity of the concrete
$E_{s,Ernst}$	Table **4.1**	Equivalent modulus of elasticity of the stay cable
γ_c	26 kN/m^3	Density of the concrete
γ_s	78 kN/m^3	Density of the steel cable
A_{c1}	20.8 m^2	Cross-sectional area of the girder
A_{c2}	26.8 m^2	Cross-sectional area of the pylon
A_s	Table **4.2**	Cross-sectional area of the cable
f_s	1770 MPa	Initial ultimate strength of the steel cable
$f_{ck,d}$	23.1 MPa	Design value of the ultimate compressive strength of the concrete
I_1	18.598 m^4	Inertia moment of the girder
I_2	118.4 m^4	Inertia moment of the pylon
Q	63.5 kN.m^{-1}	Vehicle load on the girders in mid-span

4.5.2 Deterministic Analysis

Deterministic analysis of the bridge was conducted via a finite element model in ANSYS, as shown in Figure 4.12, where the cables were simulated by LINK180 elements and the girders and pylons were simulated by BEAM188 elements. The traffic load is uniformly distributed forces on the mid-span girders. The critical failure components were the longest cables according to Liu et al. (2016). Taking the C_m^{31}, C_m^{32}, and C_m^{33} cables and the P_1, P_2, and P_3 girders as examples, their dynamic responses

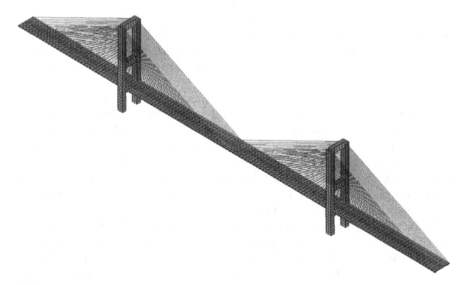

FIGURE 4.12 Finite element model of the cable-stayed bridge.

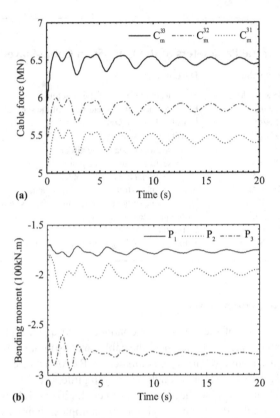

(a)

(b)

FIGURE 4.13 Response histories of the critical points subjected to sudden rupture of the $C_m{}^{34}$ cable: (a) cable force; (b) bending moment.

under the sudden failure of the $C_m{}^{34}$ cable are shown in Figure 4.13. As observed from Figure 4.13, both the cable forces and the bending moments increase rapidly following the sudden rupture of a stay cable. This phenomenon leads to the vibration of these components, which weaken over time.

Comparisons between the static and dynamic increase rates of the critical components (indicated as δ_s and δ_d, respectively) due to the cable $C_m{}^{34}$ failure are summarized in Table 4.3. It is observed that the component close to the ruptured cables has a larger increase rate. The static increase rate for the cable force ($C_m{}^{33}$) and the maximum bending moment (P_3) are 7.12% and 6.46%, respectively. The dynamic increase rate for the maximum cable force and the maximum bending moment of the girder are 11.10% and 11.79%, respectively. This phenomenon indicates that the dynamic effect due to a cable failure cannot be ignored.

Before utilizing the response data to approximate response functions, a parametric study was conducted to identify the most sensitive parameters with respect to the maximum response. The parameters in Table 4.1, except for the resistance terms of f_s and f_{ck}, were selected for the parametric sensitivity study. The sensitivities were evaluated by the following steps: (1) evaluate the response of a bridge element (defined as a_0) considering the mean value of all random parameters; (2) reevaluate the bridge

TABLE 4.3

Increase Rates of Critical Internal Forces Subjected to $C_m{}^{34}$ Rupture

Component	$C_m{}^{33}$	$C_m{}^{32}$	$C_m{}^{31}$	P_1	P_2	P_3
δ_s	7.12%	6.55%	6.05%	3.53%	4.84%	6.46%
δ_d	11.10%	9.29%	8.01%	7.65%	11.05%	11.79%

response (defined as a_i) under the condition that the mean value of the ith random variable increases by 10%.; (3) repeat the second step to compute each a_i; (4) the sensitivity for the ith random variable is $s_i = \dfrac{a_i - a_0}{\displaystyle\sum_{i=1}^{n}\left(a_i - a_0\right)^2}$. The sensitivity of each parameter with respect to the cable force and the bending moment of the girder is shown in Figure 4.14.

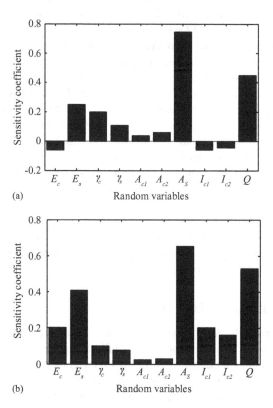

FIGURES 4.14 Sensitivity of structural parameter on: (a) $C_m{}^{33}$ cable force; (b) P_3 bending moment.

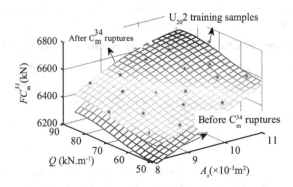

FIGURE 4.15 Response surfaces of the C_m^{33} cable force subjected to the C_m^{34} rupture.

It can be seen that the most sensitive parameters are A_s and Q, followed by E_s and γ_c. Therefore, the four parameters were selected as random variables for structural reliability analysis. In addition, A_s and E_s values were assigned following a lognormal distribution with a coefficient of variation (COV) of 0.05; γ_c was assigned following a normal distribution with a COV = 0.1; and Q was assigned following a Gumbel distribution with a COV = 0.1.

In the learning machine, the input data are the uniformly distributed samples composed of A_s, E_s, γ_c and Q, and the output data are the maximum value in the response history, as shown in Figure 4.15. A uniformly design scheme with 20 samples in a DPS program (Tang and Zhang 2013) was adopted. The response surfaces of the critical components in the initial structure or the updated structure were approximated via the learning machine considering dynamic effects. Figure 4.13 shows the response surfaces of the C_m^{34} cable force, where Q and A_s were considered as variables and E_s and γ_c were considered as constants.

It was observed that the training samples are uniformly distributed in the coordinate space, and the response surface fits the training samples fairly well. The response surface is nonlinear because of the dynamic effect and the nonlinear behavior of the structure. In addition, the C_m^{34} failure leads to an evident shift of the response surface, which has been captured by the support vector updating. With the explicit response surfaces, the component reliability can be evaluated via the first-order reliability method or the MCS approach.

4.5.3 SYSTEM RELIABILITY EVALUATION

The system failure of the cable-stayed bridge was considered with bending failure of any girders or pylons. Therefore, the bending failure at points P_1 and P_2 on the pylon and the bending failure at point P_3 on the girder were considered aspects of a system failure. Since there are 34 pairs of cables on each side of the pylons and they are highly correlated, only the longest cables of C_m^{34} and C_s^{34} were selected as the first layer of the event tree. Two event trees for $T = 0$ year and $T = 20$ years were constructed as shown in Figure 4.16 (a), (b) via the β-unzipping approach.

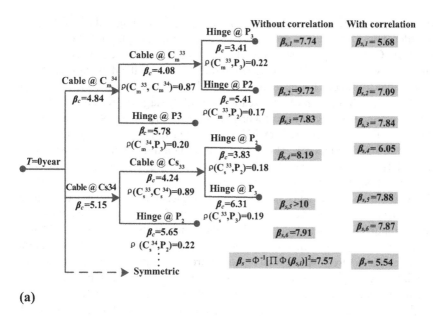

(a)

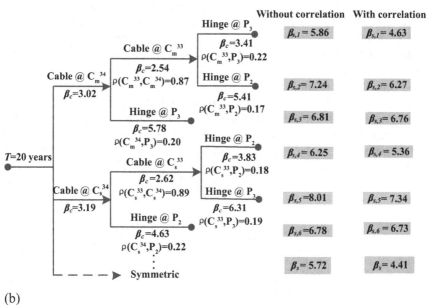

(b)

FIGURE 4.16 Three-level event trees of the long-span cable-stayed bridge at: (a) $T = 0$ year; (b) $T = 20$ years.

In Figure 4.16 (a), (b), β_c is a component reliability index evaluated from the corresponding SVR model, $\beta_{s,i}$ is a system reliability index of the ith failure path with β_c in parallel, β_s is the system reliability index that is composed of $\beta_{s,i}$ in series, and ρ is the correlation coefficient between two failure components. Two dominant failure paths were observed from the event tree. The first failure path is the strength failure of the mid-span cables (C_m^{34} and C_m^{33}) followed by the bending failure of the mid-span girder (Hinge @ P_3). The second failure path is the strength failure of the side-span cables (C_s^{34} and C_s^{33}) followed by the bending failure of the pylons (Hinge @ P_2). In this case study, the cable degradation did not change the domain failure path over the cable lifetime. This phenomenon may be explained by the fact that long-span cable-stayed bridges have dense cables providing large degrees of redundancy that can avoid the bending failure of a bridge girder. Therefore, cable degradation or a cable loss will not change the domain failure mode from a cable to a girder failure. It can also be inferred that the cable strength degradation has a greater effect on the system safety of a cable-stayed bridge with long-spacing (e.g., 30 m) cables compared to that of short-spacing (e.g., 6 m) cables.

The correlation coefficients in the event trees were evaluated based on their response surface functions. According to the concept of the probability network estimating technique (PNET) (Ang et al. 1975), the components can be classified based on their correlation coefficients. In general, the components with correlation coefficients larger than a threshold coefficient, i.e., approximately 0.8 (Liu et al. 2014), can be simplified as a representative component; otherwise, the two components can be considered independently. In the first failure path shown in Figure 4.16 (a), ρ (C_m^{34} and C_m^{33}) = 0.87, higher than the demarcating coefficient; ρ (C_m^{34} and C_m^{33}) = 0.22, lower than the demarcating coefficient. As a result, the consideration of the correlation effect results in a significant decrease of the reliability $\beta_{s,1}$ from 7.76 to 5.68. This phenomenon results from the highly correlated cable strength and response models which reduce the number of independent components in a parallel failure path. Thus, if correlation between two failure modes is not considered it will provide a non-conservative result.

The target system reliability index β_T is chosen herein as a reference to evaluate the scheme for cable replacement. The MOHURD in China (1999) recommends β_T = 5.2, and the AASHTO (2004) recommends β_T = 5.0 ~ 6.0, as suggested by Nowak and Szerszen (2000). The lifetime system reliability of the cable-stayed bridge considering correlation was reevaluated as shown in Figure 4.17, where T_L is the design lifetime of a cable. As observed in Figure 4.17, the service times of the bridge

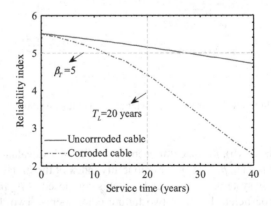

FIGURE 4.17 System reliability of the cable-stayed bridge subjected to cable degradation.

corresponding to $\beta_T = 5.0$ for bridges with uncorroded and corroded cables are approximately 28 and 12 years, respectively.

It can be inferred from the case study that the cable failures in the event tree are highly correlated, contributing to an evident decrease in the system reliability. Although the high degree of redundancy in a cable-stayed bridge contributes to the system safety of a cable-stayed bridge, highly corroded and correlated cables will reduce its system safety. Therefore, the consideration of cable corrosion and correlation is important for lifetime safety evaluation of in-service cable-stayed bridges. A further analysis based on the proposed computational framework and site-specific cable inspection data could provide a theoretical basis for cable replacement.

4.6 CONCLUSIONS

A computational framework is proposed to evaluate the system reliability of cable-stayed bridges subjected to cable degradation. This framework can be applied as a general tool to any cable-supported bridge considering cable degradation. It advances the state of the art by capturing the influence of cable strength reduction on bridge system reliability over its lifetime. The effects of corrosion and correlation between defects are considered in the strength reduction model.

The case studies of two cable-stayed bridges, including an ancient short-span bridge with long cable spacing (30 m) and a modern long-span bridge with short cable spacing (6 m), both demonstrate the influence of the cable degradation on the structural system reliability. Several critical factors affecting the bridge lifetime safety were found from the case studies. The conclusions are summarized as follows.

1. The cable strength reduction due to fatigue-corrosion effect over the cable lifetime can be up to 25%, and the dynamic amplification factor of the load effect resulting from a cable rupture can be as high as 11%. These features and the structural nonlinear behavior have been captured by updating the learning machines in the proposed computational framework.
2. The cable strength degradation due to the fatigue-corrosion effect can change the dominant failure path of a cable-stay bridge, thereby leading to a significant reduction in the bridge system reliability. This phenomenon is associated with cable spacing, where a sparse cable system appears to be more sensitive to cable strength reduction.
3. In addition to the cable degradation, the strong correlation in cable defects also leads to a reduction in system safety. Although a cable-stayed bridge has high degrees of redundancy that contribute to the system safety, strongly corroded cables with high correlation in the damage reduces the system safety of the bridge.
4. For a target system reliability index of 5.0, the corresponding lifetime of the long-span cable-stayed bridge is 18 years. In order to extend this lifetime, the proposed framework provides a basis for targeted cable replacement in conjunction with site-specific inspection data.

A better understanding of the cable degradation effect on the system reliability of cable-stayed bridges was provided by this study. However, the system reliability evaluation of cable-stayed bridges is a comprehensive task and can be improved further in the future: (1) a further analysis based on site-specific cable inspection data will be conducted to provide a theoretical basis for deriving cable replacement schemes; (2) a complete event tree of the cable-stay bridge will be built for an in-depth investigation; and (3) the influence of the cable spacing on the structural system reliability deserves further investigation.

REFERENCES

AASHTO. (2004). *AASHTO LRFD bridge design specifications*, Washington, DC.

Alibrandi, U., Alani, A. M. and Ricciardi, G. 2015. A new sampling strategy for SVM-based response surface for structural reliability analysis. *Probabilistic Engineering Mechanics* 41: 1–12.

ANSYS [Computer software]. n.d. *ANSYS*, Canonsburg, PA, USA.

Ang, H. S., Abdelnour, J. and Chaker, A. A.. 1975. Analysis of activity networks under uncertainty. *Journal of the Engineering Mechanics Division* 101: 373–387.

Aoki, Y., Valipour, H. Samali, B. and Saleh, A. 2013. A study on potential progressive collapse responses of cable-stayed bridges. *Advances in Structural Engineering* 16: 689–706.

Bruneau, M. 1992. Evaluation of system-reliability methods for cable-stayed bridge design. *Journal of Structural Engineering* 118: 1106–1120.

Chang, C. C. and Lin, C. J. 2011. LIBSVM: a library for support vector machines. *ACM Transactions on Intelligent Systems and Technology*.

Cheng, J. and Xiao, R. C. 2005. Serviceability reliability analysis of cable-stayed bridges. *Structural Engineering & Mechanics* 20: 609–630.

CSRA [Computer software]. 2013. *Complex Structural Reliability Analysis V1.0*, Changsha University of Science and Technology, Changsha, China.

Dai, H., Zhang, H. and Wang, W. 2012. A support vector density-based importance sampling for reliability assessment. *Reliability Engineering and System Safety* 106: 86–93.

Daniels, H. 1945. The statistical theory of the strength of bundles of threads, Part I. *Proceedings of the Royal Society A Mathematical, Physical and Engineering Sciences* 183: 405–435.

Estes, A. C. and Frangopol, D. M.. 2001. Bridge lifetime system reliability under multiple limit states. *Journal of Bridge Engineering* 6: 523–528.

Faber, M. H., Engelund, S. and Rackwitz, R. 2003. Aspects of parallel wire cable reliability. *Structural Safety* 25: 201–225.

Freire, A. M. S., Negrao, J. H. O. and Lopes, A. V. 2006. Geometrical nonlinearities on the static analysis of highly flexible steel cable-stayed bridges. *Computes and Structures* 84: 2128–2140.

Jia, B., Yu, X. Yan, Q. and Zhen, Y. 2016. Study on the system reliability of steel-concrete composite beam cable-stayed bridge. *The Open Civil Engineering Journal* 10: 418–432

Kim, D. S., Ok, S. Y. Song, J. and Koh, H. M. 2013. System reliability analysis using dominant failure modes identified by selective searching technique. *Reliability Engineering System Safety* 119: 316–331.

Lee, Y. J. and Song, J. 2011. Risk analysis of fatigue-induced sequential failures by branch-and-bound method employing system reliability bounds." *Journal of Engineering Mechanics* 137: 807–821.

Li, H., Li, S. Ou, J. and Li, H. 2012. Reliability assessment of cable-stayed bridges based on structural health monitoring techniques. *Structure and Infrastructure Engineering* 8: 829–845.

Li, H., Lan, C. M., Ju, Y. and Li, D. S.. 2012. Experimental and numerical study of the fatigue properties of corroded parallel wire cables. *Journal Bridge Engineering* 17: 211–220.

Li, S., Xu, Y. Li, H. and Guan, X. 2014. Uniform and pitting corrosion modelling for high strength bridge wires. *Journal of Bridge Engineering* 19: 04014025.

Li, S., Xu, Y. Zhu, S. Guan, X. and Bao, Y. 2015. Probabilistic deterioration model of high-strength steel wires and its application to bridge cables. *Structure & Infrastructure Engineering* 11(9): 1240–1249.

Li, W., Yan, Q. S. and Wang, W. 2010. System reliability analysis of cable-stayed bridge. *Journal of Shenyang University of Technology* 32: 235–240. (in Chinese)

Liu, Y., Lu, N. Noori, M. and Yin, X. 2014. System reliability-based optimization for truss structures using genetic algorithm and neural network. *International Journal of Reliability and Safety* 8: 51–69.

Liu, Y., Lu, N. Yin, N. and Noori, M. 2016. An adaptive support vector regression method for structural system reliability assessment and its application to a cable-stayed bridge. *Proceedings of the Institution of Mechanical Engineers Part O Journal of Risk & Reliability* 230: 204–219.

Liu, Y., Lu, N. and Yin, X. 2016. A hybrid method for structural system reliability-based design optimization and its application to trusses. *Quality and Reliability Engineering International* 32: 595–608.

Lu, W. and He, Z. 2016. Vulnerability and robustness of corroded large-span cable-stayed bridges under marine environment. *Journal of Performance of Constructed Facilities* 30: 04014204.

Lu, N., Liu, Y. and Beer, M. (2018). System reliability evaluation of in-service cable-stayed bridges subject to cable degradation. *Structure and Infrastructure Engineering* 14(11): 1486–1498.

Mahmoud, K. M. 2007. Fracture strength for a high strength steel bridge cable wire with a surface crack. *Theoretical & Applied Fracture Mechanics* 48: 152–160.

Maljaars, J. and Vrouwenvelder, T. 2014. Fatigue failure analysis of stay cables with initial defects: Ewijk bridge case study. *Structural Safety* 51: 47–56.

MATLAB R 2010 [Computing software]. *MathWorks*, Natick, MA.

Marjanishvili, S. M. 2004. Progressive analysis procedure for progressive collapse. *Journal of Performance of Constructed Facilities* 18: 79–85.

Mehrabi, A. B., Ligozio, C. A. Ciolko, A. T. and Wyatt, S. T. 2010. Evaluation, rehabilitation planning, and stay-cable replacement design for the Hale Boggs bridge in Luling, Louisiana. *Journal of Bridge Engineering* 15: 364–372.

MOCAT. (2010). *JT/T 775–2010: Stay cables of parallel steel wires for large-span cable-stayed bridge*, Ministry of Communications and Transportation, Beijing, China.

MOHURD. (1999). *GB/T 50283–1999: Unified standard for reliability design of highway engineering structures*, Ministry of Housing and Urban-Rural Development, Beijing, China.

Mozos, C. M. and Aparicio, A. C.. 2011. Numerical and experimental study on the interaction cable structure during the failure of a stay in a cable stayed bridge. *Engineering Structures* 33: 2330–2341.

Nakamura, S. and Suzumura, K. 2012. Experimental study on fatigue strength of corroded bridge wires. *Journal of Bridge Engineering* 18: 200–209.

Nariman, N. A.. 2017. Aerodynamic stability parameters optimization and global sensitivity analysis for a cable stayed bridge. *KSCE Journal of Civil Engineering* 21: 1–16.

Nowak, A. S. and Szerszen, M. M. 2000. Structural reliability as applied to highway bridges. *Progress in Structural Engineering & Materials* 2: 218–224.

Starossek, U. 2007. Typology of progressive collapse. *Engineering Structures* 29: 2302–2307.

Stewart, M. G. and Al-Harthy, A. 2008. Pitting corrosion and structural reliability of corroding RC structures: Experimental data and probabilistic analysis. *Reliability Engineering & System Safety* 93: 373–382.

Thoft-Christensen, P. and Murotsu, Y. 1986. *Application of structural systems reliability theory*, Speinger-Verlag Berlin, Heidelberg.

Tang, Q. Y. and Zhang, C. X.. 2013. Data Processing System (DPS) software with experimental design, statistical analysis and data mining developed for use in entomological research. *Insect Science* 20: 254–260.

Weibull, W. 1949. *A statistical representation of fatigue failure in solids*. Transactions on the Royal Institute of Technology. Stockholm, Sweden.

Wolff, M. and Starossek, U. 2009. Cable loss and progressive collapse in cable-stayed bridges. *Bridge Structures* 5: 7–28.

Xu, J. and Chen, W. 2013. Behavior of wires in parallel wire stayed cable under general corrosion effects. *Journal of Constructional Steel Research* 85: 40–47.

Xu, Y., Li, H. Li, S. Guan, X. and Lan, C. 2016. 3-d modelling and statistical properties of surface pits of corroded wire based on image processing technique. *Corrosion Science* 111: 275–287.

Yang, W., Yang, P. Li, X. and Feng, W. 2012. Influence of tensile stress on corrosion behaviour of high-strength galvanized steel bridge wires in simulated acid rain." *Materials and Corrosion* 63: 401–407.

Yang, O., Li, H., Ou, J. and Li, Q. S. 2013. Failure patterns and ultimate load-carrying capacity evolution of a prestressed concrete cable-stayed bridge: case study. *Advances in Structural Engineering* 16: 1283–1296.

Yoo, H., Na, H. S. Choi, E. S. and Choi, D. H.. 2010. Stability evaluation of steel girder members in long-span cable-stayed bridges by member-based stability concept. *International Journal of Steel Structures* 10: 395–410.

Yoo, H., Na, H. S. and Choi, D. H.. 2012. Approximate method for estimation of collapse loads of steel cable-stayed bridges. *Journal of Constructional Steel Research* 72: 143–154.

Zhou, Y. and Chen, S. 2014. Time-progressive dynamic assessment of abrupt cable-breakage events on cable-stayed bridges. *Journal of Bridge Engineering* 19: 159–171.

Zhou, Y. and Chen, S. 2015. Numerical investigation of cable breakage events on long-span cable-stayed bridges under stochastic traffic and wind. *Engineering Structures* 105: 299–315.

Zhu, J. and Wu, J. 2011. Study on system reliability updating through inspection information for existing cable-stayed bridges. *Advanced Materials Research* 250–253, 2011–2015.

5 Reliability Evaluation of a Cable-stayed Bridge Subjected to Cable Rupture During Construction

Naiwei Lu

Changsha University of Science and Technology, China

Yang Liu

Hunan University of Technology, China

CONTENTS

5.1   INTRODUCTION

Due to their structural long-span capability and economic advantages cable-stayed bridges are used widely in highway networks to cross large canyons and rivers (Gimsing and Georgakis 2011). A cable-stayed bridge is a complex system composed of girders, cables, pylons, and piers, which keeps the structural integrity. Cables are critical components connecting girders and pylons and transmitting the structural dead load and external live loads. The cables of cable-stayed bridges have a higher risk of failure during construction due to human error and structural uncertainties (Altunisik et al. 2010; Atmaca and Ates 2012). For instance, nine stay cables on Chishi Bridge in China suddenly ruptured as a result of inappropriate construction measures which led to a large number of cracks on the bridge girder (Wiki, 2014). As a result, the cantilever girder of the bridge fell by 2 m and several large cracks were observed in the concrete bridges. Choi et al. (2006) indicated that the main reason for the collapse of Haengju Grand Bridge in Korea was the rupture of cables. A load case of the cable loss may lead to failure of more cables and even failure of the whole structure. Recommendations for robustness design and progressive collapse design are highlighted in the design codes of cable-stayed bridges. For instance, for evaluating the response of cable-stayed bridges to the sudden loss of a cable, PTI (2012) suggest a static analysis with a dynamic amplification factor of 2.0 or a nonlinear dynamic analysis with no less than 1.5 times the static force. However, the uncertainties which exist during construction make the cable more vulnerable. The cable rupture is more likely to happen during construction, mainly due to uncertainties and the structural lower degrees of freedom. In addition, risks and uncertainties associated with accidental loading and structural parameters can also lead to the failure of stay cables. Therefore, special attention should be paid to the safety of cable-stayed bridges subject to failure of a cable stay in the construction stages. From the point of view of mechanical performance, the sudden rupture of a stay cable is a dynamic load leading to nonlinear vibration of the bridge. In this respect, Aoki et al. (2011) studied the impact of sudden failure of cables on the dynamic performance of a cable-stayed bridge. Lonetti and Pascuzzo (2014) investigated the vulnerability of the structure against damage and complete failure phenomena produced by the longest central system stays. The effect of the cable failure on the stress distribution in the bridge components and the corresponding dynamic influence produced by the cable release mechanisms is significant for cable-stayed bridges.

Structural uncertainties during construction further complicate the reliability evaluation of cable-stayed bridges. Most long-span bridges are statically indeterminate, while their construction states are determinate. Numerous research efforts have been made to address these uncertainties. Bruneau (1992) found a large number of potential failure sequences of a short-span cable-stayed bridge by using a branch-bound approach. Cheng and Li (2009) took the static wind load-related parameters as variables and evaluated the reliability of a steel arch bridge under wind load. Cho and Kim (2008) evaluated the risks in a suspension bridge by considering an ultimate limit state for the fracture of main cable wires during construction phases. Cheng and Xiao (2005) evaluated the serviceability reliability of a cable-stayed bridge by combining response surface, finite element, and first order reliability methods. Lu et al. (2018) evaluated system reliability of a long-span cable-stayed bridge during service period taking consideration of cable strength degradation. However, little attention has been paid to dealing with the uncertainties and reliabilities of cable-stayed bridges during construction, where the cables and girders are critically important during the cantilever stage. The vulnerability of cable-stayed bridges subject to cable loss scenarios during construction has not been thoroughly investigated and recommended in the current studies and design codes. The structural safety of cable-stayed bridge caused by failure of cables merits research.

This chapter evaluated the reliability of a cable-stayed bridge subjected to cable failure during construction by investigating its mechanical behavior when subjected to cable rupture during construction. Subsequently, an intelligent algorithm combining the β-bound and response surface methods is proposed for conducting an efficient evaluation of the system reliability. Finally, the system reliability indices of the cable-stayed bridge during different construction stages are evaluated with consideration of the domain failure paths.

5.2 MECHANICAL BEHAVIOR OF A CABLE-STAYED BRIDGE SUBJECTED TO CABLE RUPTURE DURING CONSTRUCTION

5.2.1 DESCRIPTION OF A CONCRETE CABLE-STAYED BRIDGE

The prototype cable-stayed bridge is the Hejiang Yangtze River Bridge in Sichuan, China (Lu et al. 2019). Consider that several cables of the bridge were ruptured due to the welding spatter as a result of human error in construction. Figures 5.1 (a, b) show the cantilever state and the ruptured cables during construction. Dimensions of the cable-stayed bridge are shown in Figure 5.2. The bridge girders, towers, and piers are made of C60 concrete, and the cables are made of steel strands. The girders and the cables are divided into individual serial segments as shown in Figure 5.2, where CBA1 denotes the first cable element in side-span, and GNJ1 represents the first girder element in north mid-span. There are 36 construction states; 34 cantilever states and two closure states. During the first construction stage, the GBJ1, GBA1, CBJ1, and CBA1 elements were activated. The following states were implemented by installing the corresponding elements.

In the cantilever states, a pair of steel cradles was used for pouring wet concrete into the bridge girders. The weight of each steel cradle, which was a temporary load, was 124 tonnes. In addition, the rebar and its equivalents were also temporary load. The dead load was the constructed concrete girders and cables. Rupture of cables delivered dynamic load to the bridge girders and towers, which are critical for their failure. Therefore, reliability of the structural system under these loads justifies investigation.

5.2.2 CRITICAL SCENARIOS OF CABLE RUPTURE

The thirty-second cantilever state was selected as the prototype construction state. Five scenarios were identified and chosen as numerical studies, as shown in Figure 5.3. Illustrations of the symbols are shown in Table 5.1.

As shown in Figure 5.3 and Table 5.1, SC0 is the initial state without cable rupture. SC1 is the first scenario considering cable failure of CBJ32, and SC5 is the fifth scenario considering the cable failure of CBJ28 ~ CBJ32.

(a)

(b)

FIGURE 5.1 Photos of the cable-stayed bridge during construction in: (a) cantilever state; (b) cable rupture state.

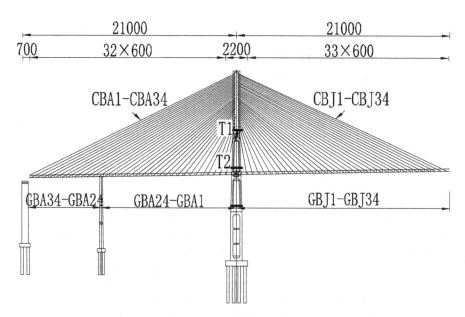

FIGURE 5.2 Dimensions of the cable-stayed bridge.

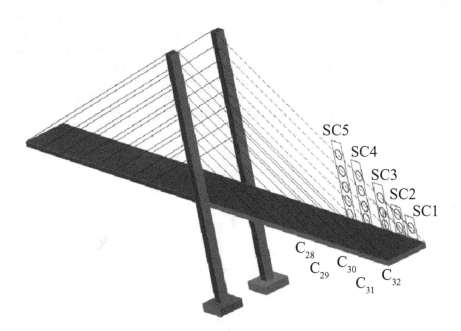

FIGURE 5.3 Scenarios of cable rupture.

TABLE 5.1
Illustration of Symbols

Symbols	Illustration
C28 ~ C32	Cables numbered 28 to 32
SC1	Scenario No. 1 with failure of C32
SC2	Scenario No. 2 with failure of C32 and C31
SC3	Scenario No. 3 with failure of C32, C31, and C30
SC4	Scenario No. 4 with failure of C32, C31, C30, and C29
SC5	Scenario No. 5 with failure of C32, C31, C30, C29, and C28

5.2.3 FINITE ELEMENT SIMULATION OF EACH SCENARIO

A finite element model shown in Figure 5.4 was built based on a commercial software package. The pylons and girders were beam elements and the cables were link elements. The temporary loads were considered as concentrated forces on the girders. The cable rupture was considered as passivized elements. The dynamic effects due to cable rupture were considered in the transient analysis.

The first step is to analysis the influence of cable rupture on the cable forces. We consider the cable forces of five cables (CBJ28 ~ CBJ32) and the stress of girders and towers under the five scenarios. In the case of SC1, delete the CBJ32 element directly and then enter the transient analysis module. Internal forces of the critical components are shown in Figures 5.5 (a, b).

As observed from Figure 5.5(a), the sudden rupture of the CBJ32 element leads to an increased rate of 4.8% and 3.2% for the cable forces of CBJ31 (the red line) and CBJ32 (the blue line). This phenomenon indicates that failure of a stay cable leads to significant force increase in the cables on the same side, rather than the opposite side. As observed from Figure 5.5(b), the sudden rupture of the CBJ32 element leads to a variation of the bending moment of the towers and girders. The girder element has a relatively larger bending moment compared with the tower elements.

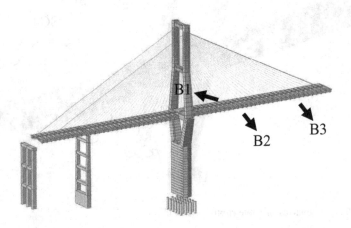

FIGURE 5.4   Finite element model of the cable-stayed bridge.

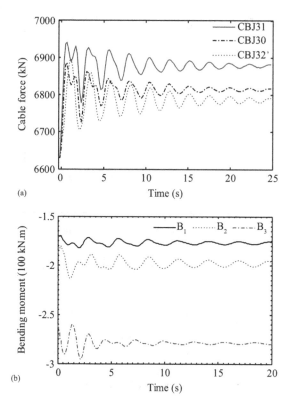

(a)

(b)

FIGURE 5.5 Internal forces of the critical components under the case of CBJ31 rupture: (a) cable forces; (b) bending moment.

Considering the five scenarios of cable rupture, the cable force and the girder stress were evaluated based on the finite element simulation. Figures 5.6 (a, b) show the variation trend of the cable force and the girder stress under different scenarios of cable rupture. Figure 5.6(a) shows that an increase in the number of raptured cables leads to an increase of cable force. A total number of five cable ruptures leads to a 74% increase in the CBJ31 cable force. As observed from Figure 5.6 (b), the cable rupture has a slight influence on the girder stress. However, the girder stress has an obvious variation, where the five cable rupture leads to the lower point stress of the GBJ16 element increasing from −2.1 MPa to −14.8 MPa.

Based on the above numerical results, it can be surmised that the critical components in the case of cable rupture are the long cables and the GBJ16 girders (1/4 span point).

5.3   A FRAMEWORK FOR SYSTEM RELIABILITY EVALUATION OF CABLE-STAYED BRIDGES SUBJECTED TO CABLE RUPTURE

5.3.1   System Properties of Cable-stayed Bridges

A cable-stayed bridge has special properties compared with short-span bridges. The cables make the structural system more complicated than girder bridges, which also

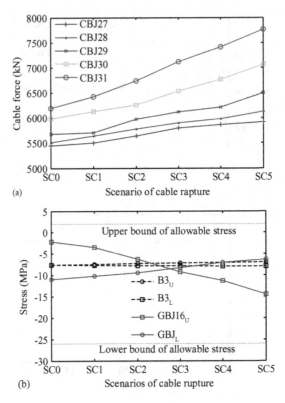

FIGURE 5.6 Influence of the scenarios of the cable rupture on the component: (a) cable force; (b) stress.

produces nonlinearity and multiple degrees of freedom. The nonlinearities, such as the cable sag effect, the beam-column effect, and large displacement, will impact the shape of the limit state function.

For the cable sag effect simulation, the common approach is Ernst's equation. A nonlinear limit state function can be established based on the numerical simulation. Regarding the beam-column effect, a second-order effect and be conveniently considered in a stability function:

$$\frac{M}{M_P} = 1 - \left(\frac{P}{P_P}\right)^2 \frac{A^2}{4wZ_x} \tag{5.1}$$

where M and M_p are the applied and plastic moment in the absence of axial loads, respectively; P and P_P are the applied and plastic axial force, respectively; w is the web thickness, and Z_x is the bending plastic modulus. Suppose the term of $\dfrac{A^2}{4wZ_x}$ on the right side of Equation (5.4) is equal to 1, and the beam-column curve is shown in Figure 5.7. It is observed from Figure 5.7 that the curve is nonlinear due to the

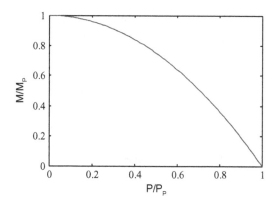

FIGURE 5.7 A beam-column effect curve.

relationship between the axle force and bending moment. The nonlinearity of the limit state function will be captured by a learning machine as illustrated below.

In addition to the component-level properties of cable-stayed bridges, the system-level behavior is another significant factor needing consideration. From a system point of view, the rupture of a stay cable will lead to a variation of the internal force of the remaining elements. The system failure can be modeled in a mathematical model composed of elements connecting in parallel or series. The reliability index of each component is actually a conditional reliability index. Suppose a system is composed of n components, and the k-1 components have failed. The reliability index of the kth failure component can be written as:

$$\beta_{n_k/}^{(k)} = \beta_{n_k/n_1,n_2,\cdots n_{k-1}}^{(k)} = \Phi^{-1}\left[P\left(E_{n_k/}^{(k)}\right)\right] \tag{5.2}$$

where, $E_{n_k/}^{(k)}$ is the event of the kth component failure, P is the probability of the event, Φ^{-1} is an inverse cumulative distribution function, and $\beta_{n_k/n_1,n_2,\cdots n_{k-1}}^{(k)}$ is a conditional reliability index of the kth potential component.

Based on the above illustration, it is clear that the nonlinearity of the structural mechanical behavior will lead to the time-consuming problem of a reliability computation. Furthermore, system updating will lead to more computations. Thus, it is necessary to utilize an effective computational framework to estimate the system reliability of a cable-stayed bridge.

5.3.2 AN EFFECTIVE COMPUTATIONAL FRAMEWORK FOR STRUCTURAL SYSTEM RELIABILITY EVALUATION

An adaptive machine learning approach is presented herein to evaluate the system reliability of the stay cables during construction. The flowchart of the framework is shown in Figure 5.8. Compared with the traditional machine learning approach, there are two updating approaches; the first updating procedure searches the design point, and the second updating procedure searches the failure path.

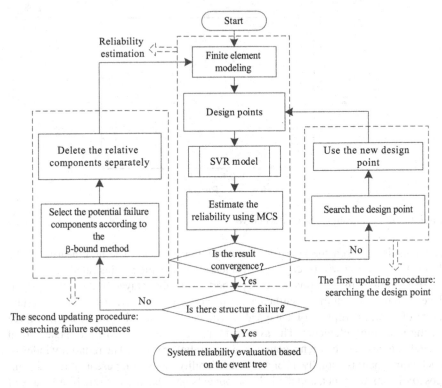

FIGURE 5.8 Flowchart of the computational framework for system reliability evaluation of cable-stayed bridges.

In the first updating procedure, a machine learning scheme of support vector machine is utilized to approximate the structural limit state function. A linear function with nonlinear kernel functions combined in a high-dimension feature space is used to capture the structural nonlinear behavior. The formula of the learning machine is written as:

$$g(x) = w \cdot \varphi(x) + b \tag{5.3}$$

where, the $\varphi()$ is a kernel function, w denotes the weight vector in primal weight space, and b is the bias term. With the structural risk minimization (SRM) principle, the SVR is formulated as minimization of the following function:

$$\min \quad J(w,e) = \frac{1}{2}w^2 + \frac{1}{2}C\sum_{i=1}^{L}e_i^2 \tag{5.4}$$

$$\text{s.t.} \quad y_i = w^T\varphi(x_i) + b + e_i \qquad (i = 1,2,\ldots,L)$$

where C is a regularization constant, and e is an error vector. In order to solve the optimum problem shown in Equation (5.4), a Lagrange Multiplier optimal programming method can be adopted. Finally, the SVR model for function estimation is represented as

$$f(x) = \sum_{i=1}^{l} \alpha_i \psi(x, x_i) + b \tag{5.5}$$

where $\psi(x, x_i)$ is the kernel function which enables the dot product to be computed in a high-dimension feature space using low-dimension space data input without the transfer functions. The kernel function in this paper is a Gaussian function.

The purpose of the second updating procedure is to evaluate the failure paths of the bridge system. When evaluating the reliability index of the component based on the first updating procedure, the relative component will be deleted from the initial finite element model and the system will then be updated. In order to improve the efficiency of the updating procedure, a β-bound method is utilized. A structural system failure event E_s is made of m independent failure modes $E_i(i = 1,...,m)$, where E_i includes some failure events $E_i^j(j = 1,...,n)$ in sequences. The structural system reliability is obtained by

$$\begin{cases} E_i = \bigcap_{j=1}^{n} E_i^j \\ E_s = \bigcup_{i=1}^{m} E_i \end{cases} \tag{5.6}$$

As shown in Equation (5.6), each single failure event is composed of multiple failure states in series, and the system reliability failure events are composed of multiple failure events in parallel. Accordingly, the entire structure can be modeled as a series–parallel combination of failure modes. The system reliability is estimated on the basis of failure sequence models. The benefit of the second procedure is that the updating procedure is computationally efficient. Furthermore, the updating procedures associated with the SVR model enable the excellent SVR approach to be used in a structural fault event tree.

5.4   SYSTEM RELIABILITY EVALUATION OF A CABLE-STAYED BRIDGE DURING CONSTRUCTION

5.4.1   RANDOM VARIABLES AND SENSITIVE ANALYSIS

In order to evaluate the system reliability of the Hejiang Bridge shown in Figure 5.2 during the construction period, a total number of 12 random variables were selected

TABLE 5.2

Statistical Distribution of the Random Variables

Variables	Variables (unit)	Distribution	Mean	Standard deviation
Elastic modules of girders	E_1	Normal	3.25E4	3.33E3
Elastic modules of pylons	E_2	Normal	3.11E4	3.35E3
Elastic modules of cable	E_3	Normal	1.95E5	1.95E4
Cross-sectional area of girder	A_1	Lognormal	20.574	1.011
Cross-sectional area of girder	A_3	Lognormal	26.324	1.524
Cross-sectional area of a stay cable	A_5	Lognormal	1.4E-4	7.0E-6
Density of girder	γ_1	Normal	26.57	1.54
Density of pylon	γ_2	Normal	26.32	1.31
Density of cable	γ_3	Normal	78.5	3.93
Moment of inertia of girder	I_1	Lognormal	18.498	0.930
Moment of inertia of pylon	I_3	Lognormal	118.432	5.725
Temporary load	Q	Gumbel	See Equation (5.7)	

for the present study. The random variables and their statistical characteristics of a cable-stayed bridge are shown in Table 5.2.

The probabilistic model of the temporary load is referred to Liu et al. (2004), written as:

$$
\begin{cases}
\mu_Q = 769.7e^{-0.7848(t-(t-\mathrm{mod}(t,d)+6)/6+3} \\
\sigma_Q = -21.237\mathrm{Ln}\left(i-\left(t-\mathrm{mod}(t,6)+6\right)+3\right)+32.454
\end{cases}
\tag{5.7}
$$

where, μ_Q are σ_Q are the mean value and the standard deviation of the temporary load of the bridge during construction, respectively; t is the time, mod() is a cofunction. The resistance of the concrete will increase with the construction time. The PDF of the section distribution of the concrete can be written as:

$$
f_R(x,t) = \frac{1}{\sqrt{2\pi}\sigma_R(t)x}\exp\left\{-\frac{\ln x - \mu_R(t)}{2\sigma_R^2(t)}\right\}
\tag{5.8}
$$

where $\mu_R(t)$ and $\sigma_R(t)$ are the mean value and standard deviation of the resistance of the concrete. The standard deviation of the resistance is 0.12 according to Liu et al. (2004). The resistances between these components are highly correlated.

A parametric study was conducted aimed at determining the most sensitive structural parameters with respect to the structural response. Initially, the influence of the random variables on the bending moment of the GBJ16 was analyzed. Figure 5.9 shows the normalized influence factor of the random variables. It can be seen that temporary load (Q) and the cross-sectional area of the cable (A_3) are the most sensitive factors. In addition to the two most sensitive factors, the elastic modules of the girders and pylons were also selected as random variables for the reliability evaluation.

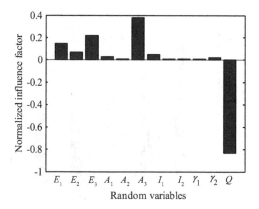

FIGURE 5.9 Influence of the random variables on the bending moment of GBJ16.

Subsequently, the influence of the cable failure on the bending moment of girders was investigated. Figure 5.10 shows the sensitive factor of the bending moment due to cable failure. It is observed that side-span cable failure is sensitive to the bending moment of the girder, while cables in the mid-span are sensitive to the bending moment of the pylon.

5.4.2 LIMIT STATE FUNCTION

It is necessary to define the notations of failure events for the purpose of conducting the system reliability evaluation. According to engineering practices, the common failure events for cable-stayed bridges are associated with the strength of cables, girders, and towers. In this study, the failure events are described as follows:

1. E_{cable}: Rupture of a cable, which is the most critical component in this study;
2. E_{girder}: Fracture of a girder due to negative bending moment;
3. E_{pylon}: Fracture of a pylon due to bending moment.

For the bending failure of the girders and pylons, the limit state function can be written as:

$$Z_i = 1 - \frac{P^i(X)}{P_u^i} - \frac{M^i(X)}{M_u^i} \quad (i = 1,\cdots,m) \tag{5.9}$$

where $P^i(X)$ and $M^i(X)$ are the axle force and the bending moment of the ith element, respectively; P_u^i and M_u^i are the capability of the axle force and the bending moment of the ith element. For the failure mode of cable rupture the limit state function of the cable rupture can be written as:

$$Z_j = T_u^j - T^j(X). \tag{5.10}$$

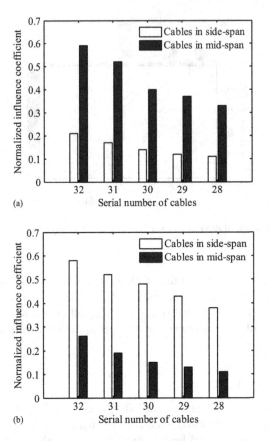

FIGURE 5.10 Influence of cables on bending moment of (a) tower; and (b) girder.

where X is the random variables associated with Q, A_3, E_1, and E3, and $T_j(X)$ is the cable force of the jth cable element.

The response surface of the bending moment of the GBJ16 element was approximated by the SVR approach. In addition, the response surface was updated considering the cable failure of the CBJ32 element. The approximated and updated response surfaces of the cable force response surface are shown in Figure 5.11.

Figure 5.11 shows that the response surface of the girder increases obviously under the case of CBJ32 rupture. This phenomenon indicates that the system updating has been captured by the SVR approach.

5.4.3 SYSTEM RELIABILITY EVALUATION

Based on the above information, the system reliability of the cable-stayed bridge during construction can be evaluated via the proposed computational framework. The system failure in the case study is considered as the bending failure of any girders or pylons. Two event trees of the bridge in different service periods were evaluated as

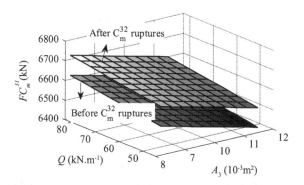

FIGURE 5.11 Influence of the failure of CBJ34 on bending moment of GBJ17.

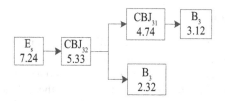

FIGURE 5.12 A two-layer failure sequence of the cable-stayed bridge during the stage of construction of the longest cantilever.

shown in Figures 5.12 (a, b) via the β-unzipping approach in conjunction with the SVR approach.

It is observed from Figures 5.12 (a, b) that there are two domain failure paths for the structural system of the cable-stayed bridge. The first domain failure path is the cable failure followed by the girder bending failure of the GBJ16 element. The second domain failure path is the cable failure followed by another cable failure, and the final failure mode is the bending failure of the B3 element. Based on the fault tree of the system, the system reliability is 7.24, which is estimated considering a series–parallel relationship.

The above analysis was conducted considering the largest cantilever state of the bridge during construction. Let us consider the next construction stat, which is the closure state of the side-span. In this construction state, the side-span is connected to the bridge pier, and the structural system has been changed from a static determinate to an indeterminate structure. In this case, the fault tree of the system is shown in Figure 5.13. It is observed that the dominant failure path of this state is the bending failure of girders.

From the case study of the system reliability evaluation, it can be concluded that the most sensitive failure mode for the cable-stayed bridge during construction is the bending failure of the bridge followed by the sudden rupture of cables. The structural system reliability decreased with the increase of cantilever girder segments, but it increases greatly since the closure of the side-span because of the conversion of a determinate to an indeterminate structure. The structural system reliability was

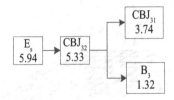

FIGURE 5.13 A two-layer failure sequence of the cable-stayed bridge during the construction of the closure of the side-span.

greatly improved since the closure of mid-span because of the increase of the degree of indeterminacy. In addition, it also demonstrates that the computational framework is effective for system reliability evaluation of bridges during construction.

5.5 CONCLUSIONS

This study presents a framework for the reliability evaluation of a concrete cable-stayed bridge during construction. Initially, the mechanical behavior of a cable-stayed bridge subjected to cable rupture during construction was investigated. Subsequently, an intelligent algorithm combining the β-bound method and the response surface method was proposed for conducting an efficient evaluation of the system reliability. Finally, the system reliability indices of the cable-stayed bridge during different construction states were evaluated with the consideration of the domain failure paths.

The numerical analysis results indicate that the main failure sequence of cable-stayed bridges during the largest cantilever stage is the failure of outside cables followed by the bending failure of mid-span girders; the main failure sequence during the closure stage of the side-span is the partial failure of the mid-span cables followed by the bending failure of mid-span girders; the structural system reliability decreased with the extend of cantilever girders, but increases greatly since the closure of the side-span because of the conversion of determinate structure and indeterminate structure; the structural system reliability improved greatly since the closure of mid-span because of the increase of the degree of indeterminacy.

There are some challenges needed for further study. First, site-specific measurements of cable strength should be used to establish a more reasonable cable strength model. Second, a more complete event tree of the cable-stay bridge rather than taking the longest cables as the first order as adopted in the present study is to be formulated. Finally, correlation coefficients will be considered to make the estimation of the system reliability more accurate.

REFERENCES

Aoki, Y., Samali, B. Saleh, A. and Valipour, H. 2011. *Impact of sudden failure of cables on the dynamic performance of a cable-stayed bridge. In Austroads Bridge Conference 8th,* Sydney, New South Wales, Australia (No. AP-G90/11).

Altunisik, A. C., Bayraktar, A. and Sevim, B. et al. 2010. Construction stage analysis of Kömürhan Highway Bridge using time dependent material properties. *Structural Engineering and Mechanics* 36(2): 207–223.

Atmaca, B. and Ates, S. 2012. Construction stage analysis of three-dimensional cable-stayed bridges. *Steel and Composite Structures* 12(5): 413–426.

Bruneau, M. 1992. Evaluation of system-reliability methods for cable-stayed bridge design. *Journal of Structural Engineering* 118(4): 1106–1120.

Choi, H. H., Lee, S. Y. Choi, I. Y. Cho, H. N. and Mahadevan, S. 2006. Reliability-based failure cause assessment of collapsed bridge during construction. *Reliability Engineering & System Safety* 91(6): 674–688.

Cheng, J. and Li, Q. 2009. Reliability analysis of a long span steel arch bridge against wind-induced stability failure during construction. *Journal of Constructional Steel Research* 65(3): 552–558.

Cho, T. and Kim, T. S.. 2008. Probabilistic risk assessment for the construction phases of a bridge construction based on finite element analysis. *Finite Elements in Analysis and Design* 44(6): 383–400.

Cheng, J. and Xiao, R. C.. 2005. Serviceability reliability analysis of cable-stayed bridges. *Structural Engineering and Mechanics* 20(6): 609–630.

Gimsing, N. J. and Georgakis, C. T.. 2011. *Cable supported bridges: concept and design.* John Wiley and Sons http://highestbridges.com/wiki/index.php?title=Chishi_Bridge.

Liu, Y., Zhang, J. R. and Li, C. X.. 2004. Calculation of time-variant reliability for concrete cable-stayed bridges during construction. *China Journal of Highway and Transport* 17(3): 31–35.

Liu, Y., Lu, N. W. and Yin, X. F.. 2016. Noori Mohammad. An adaptive support vector regression method for structural system reliability assessment and its application to a cable-stayed bridge. *Proceedings of the Institution of Mechanical Engineers, Part O: Journal of Risk and Reliability* 230(2): 204–219.

Liu, Y., Lu, N. W. and Yin, X. F. 2015. A hybrid method for structural system reliability-based design optimization and its application to trusses. *Quality and Reliability Engineering International* 32(2): 595–608.

Lonetti, P. and Pascuzzo, A. 2014. Vulnerability and failure analysis of hybrid cable-stayed suspension bridges subjected to damage mechanisms. *Engineering Failure Analysis* 45: 470–495.

Lu, N. W., Liu, Y. and Beer, M. 2018. System reliability evaluation of in-service cable-stayed bridges subjected to cable degradation. *Structure and Infrastructure Engineering* 14(11): 1486–1498.

Lu N., Liu Y. and Deng Y. 2019. Fatigue reliability evaluation of orthotropic steel bridge decks based on site-specific weigh-in-motion measurements. *International Journal of Steel Structures* 19(1): 181–192.

Mozos, C. M. and Aparicio, A. C. 2011. Numerical and experimental study on the interaction cable structure during the failure of a stay in a cable stayed bridge. *Engineering Structures* 33(8): 2330–2341.

Mozos, C. M. and Aparicio, A. C. 2010a. Parametric study on the dynamic response of cable stayed bridges to the sudden failure of a stay, Part I: Bending moment acting on the deck. *Engineering Structures* 32(10): 3288–3300.

Mozos, C. M. and Aparicio, A. C. 2010b. Parametric study on the dynamic response of cable stayed bridges to the sudden failure of a stay, Part II: Bending moment acting on the pylons and stress on the stays. *Engineering Structures* 32(10): 3301–3312.

Post Tensioning Institute (PTI). 2012. Recommendations for Stay Cable Design, Testing & Installation. ISBN 9781931085311

Wiki. 2014. https://en.wikipedia.org/wiki/Chishi_Bridge

6 Fatigue Reliability Evaluation of Orthotropic Steel Bridge Decks Based on Site-specific Weigh-in-motion Measurements

Naiwei Lu

Changsha University of Science and Technology, China

Yang Liu

Hunan University of Technology, China

Yang Deng

Beijing University of Civil Engineering and Architecture, China

CONTENTS

6.1   INTRODUCTION

In general, a steel bridge is designed with enough fatigue resistance against the cyclic vehicle load (Sim et al. 2009; Dung et al. 2015). However, recent field investigations on several collapsed steel bridges (Wang et al. 2005; Biezma and Schanack, 2007; Lalthlamuana and Talukdar, 2013) indicated that the fatigue damage induced by accidentally overloaded trucks contributed to the bridge failures. Thus fast-growing traffic volumes and loads will become a safety hazard for the fatigue safety of steel bridges, especially in developing countries. Uncertainties in traffic flows add another challenge to accurately evaluating the fatigue damage accumulation, where the probability model of fatigue damage accumulation mostly depends on site-specific traffic loading. Therefore, integrating actual traffic information into the fatigue reliability evaluation of existing steel bridges is of great importance, as it will provide a more actual evaluation result and a theoretical basis for transportation management.

Numerous research efforts have concentrated on implementing structural health monitoring (SHM) data to evaluate fatigue damage accumulation in the fatigue-critical component joints of steel bridges (Frangopol et al. 2008; Ye et al. 2012; Ni et al. 2010; Deng et al. 2015). However, the most fatigue-critical components of steel bridge decks are the welded joints under the bridge deck, where placement of the strain sensors is difficult (Sim and Uang, 2012; Gokanakonda et al. 2016). In this regard, the numerical simulation approach can capture the fatigue stress characteristics of a bridge under monitored truck loads. Meanwhile, rapidly developing computer technology and traffic weigh-in-motion (WIM) technology contribute to the efficient and accurate simulation of the fatigue damage of steel bridges under truck load.

Site-specific WIM measurements are big data that can be used for statistical analysis of traffic loading (OBrien and Enright, 2013). Numerous traffic load models have been developed based on site-specific WIM data. For instance, Wang et al. (2005) utilized WIM measurements in Florida to develop a developed a live-load spectrum by combining static responses with estimated impact factors in a 3D nonlinear truck model. Zhao and Tabatabai (2012) developed a 5-axle single-unit truck model based on WIM records in Wisconsin to supplement the permit vehicle model. OBrien et al. (2010) estimated the characteristic maximum dynamic load effects of short- to medium-span bridges with extensive WIM measurements collected at five European sites. Marques et al. (2016) implemented WIM measurements of an old railway bridge in Portugal to simulate the axle load, axle spacing, and the velocity of the train. In addition to the traffic load modeling, WIM measurements along with finite element technology have also been used to evaluate fatigue reliability of steel bridges. In this regard, Guo et al. (2012) evaluated the fatigue reliability of an orthotropic steel bridge deck using a probabilistic finite element approach. In Guo's truck load model, the axle weights were fitted by lognormal distribution functions that ignored the bimodal or trimodal characters of the truck load. Subsequently, Guo and Chen (2013) demonstrated the effectiveness of integrating the site-specific measurement and the finite element (FE) model for fatigue reliability assessment of in-service steel bridges. Ye et al. (2015) conducted a sensitivity study on the influence of the bridge FE model on the estimated stress under vehicle loads. An advanced probabilistic fatigue stress analysis approach was proposed by Zhang and Au (2016) using WIM measurements.

As elaborated above, implementation of WIM measurements is an effective way to simulate stochastic traffic loading, which can be subsequently utilized for fatigue damage evaluation of existing steel bridges. However, the balance of computational efficiency and accuracy is still a bottleneck that limits the development of existing fatigue reliability assessment approaches. In Zhang's computational framework, the calculation of thousands of stress histories with respect to the daily truck volume is obviously a time-consuming problem. Since the traffic parameters (e.g., vehicle types, driving speeds, vehicle spacing, and gross vehicle weight (GVW)) are random in nature, the statistical information of all trucks is not included in this fatigue truck load model. However, to the best of the authors' knowledge, most of the relative research efforts with respect to stochastic traffic load have focused on vehicle–bridge interaction and impact factor analysis (Zhou and Chen, 2015), while little effort has been made on stochastic fatigue truck load modeling for fatigue reliability assessment of steel bridges.

This chapter aims to develop a computational framework for evaluating fatigue reliability of steel bridge decks based on site-specific WIM measurements. Initially, a stochastic truck load model was simulated based on 2-year WIM measurements of a bridge in China. Subsequently, a meta-model approximated by neural networks was utilized to substitute the traditional FE simulation. In the case study, a steel box-girder bridge was chosen as a prototype to demonstrate the effectiveness and efficiency of the computational framework. Parametric studies were conducted to provide suggestions to traffic management based on the predicted fatigue reliability index.

6.2 STOCHASTIC TRUCK LOAD MODEL SIMULATION BASED ON WIM MEASUREMENTS

In general, a truck load model is a parametrized vehicle with statistics of site-specific traffic flows. However, the traditional truck load model is inappropriate for probabilistic modeling of fatigue damage accumulation due to the deterministic configuration and parameters in the vehicle model. Therefore, a stochastic truck load model based on site-specific WIM measurements was utilized in the present study to represent the real traffic loading.

6.2.1 WIM MEASUREMENTS

A WIM system utilizes scales or pressure sensors embedded into the road pavement to measure parameters of crossing vehicles such as axle weight and driving speed (Lydon et al. 2016). In practice, WIM systems can be used for a range of tasks, such as protection and management of highways and other infrastructure investments. In the present study, WIM measurements were used to provide a statistical database for stochastic traffic flow simulation.

A bridge WIM system in Sichuan, China, was chosen as a prototype. The WIM system has been working since the bridge was opened to the public in May 2012. Details of the WIM system can be found in Liu et al. (2015). Before statistical analysis of the measurements, a filtering process was conducted in order to identify and remove invalid records from the database. These criteria to identify the invalid

TABLE 6.1
Description of the WIM Measurements of a Highway Bridge in China

Items	Values
Time duration	May, 2013 to April, 2015
ADTT	2145
Lane number	4
Maximum gross vehicle weight (t)	164

records are: (1) the GVW is greater than 30 kN; (2) the axle weights are between 5 kN and 300 kN; (3) the vehicle length is between 3 m and 20 m; and (4) the data was completed and not flagged with a system error. An overview of the filtered WIM measurements is shown in Table 6.1. It is worth noting that the maximum GVW for a 6-axle truck is 550 kN according to the traffic laws in China (Ministry of Communications and Transportation (MOCAT), 2004). However, as is indicated in Table 6.1, the number of overloaded trucks is approximately 17 in one day. The overload rate of the maximum GVW is around 200% over the threshold value.

Based on these measurements, the trucks were classified into six categories. Illustrations of the vehicle types are shown in Figure 6.1, where V1 indicates a light car, V2 to V6 indicate 2-axle to 6-axle trucks, and AW_{ij} denotes the jth axle weight of the ith vehicle type. It is observed that around 60% of the vehicles are 2-axle trucks and light cars. In addition, some 90% of heavy trucks tend to drive in the slow traffic lane, while lighter trucks prefer driving in the fast traffic lane. This phenomenon of traffic composition impacts the fatigue reliability of steel bridges.

In addition to the large-scale statistics shown in Figure 6.1, the small-scale statistics of individual trucks were also analyzed. Taking V_6 as an example, its histogram and probability density functions (PDFs) are shown in Figure 6.2, where the Gaussian mixture model (GMM), which will be illustrated in the "proposed computational framework," was utilized to approximate the PDF. The linear regression functions between the individual axle weight and the GVW are shown in Figure 6.3, where x indicates the GVW, and y indicates the axle weight. It is assumed that the individual axle weight in the tandem and tridem axles is equal. Both large- and small-scale information provide statistical parameters for modeling the traffic loading.

6.2.2 STOCHASTIC TRUCK LOAD SIMULATION

In general, a stochastic traffic flow model can be simulated with three scales including macroscopic, mesoscopic, and microscopic scales (Chen and Wu, 2011). Since the traffic flow model is to be developed for fatigue reliability assessment, only the parameters with significant contributions to structural fatigue damage were considered. According to the stress influence lines of steel bridge decks (Ji et al. 2013), which will also be demonstrated in the case study, the vehicle type and driving lane, as well as the axle weight, were chosen for modeling the stochastic truck load. The reasons for selecting these three parameters are outlined below. First, vehicle spacing was excluded, since the vehicle gap between two trucks in the same traffic lane is much bigger than the effective range of a stress influence lane. Second, both the axle

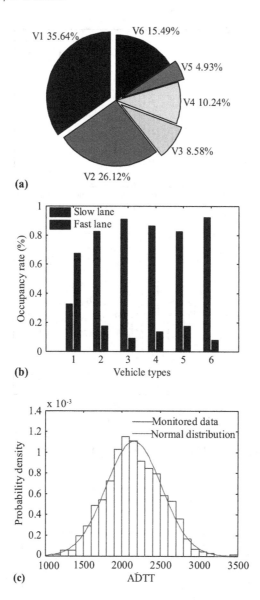

FIGURE 6.1 Probability densities of: (a) vehicle type; (b) traffic lane; (c) ADTT.

spacing and driving lane were considered, because both axle spacing and the distance between two traffic lanes are close to the effective range of a stress influence line. Finally, the driving speed was considered as a constant to take into account dynamic effects. In addition, vehicles with a GVW less than 30 kN were ignored since these vehicles contribute little to fatigue damage. As elaborated above, the stochastic fatigue truck load model is formed by three random variables including vehicle type, axle weight, and the driving lane.

With the PDF of the three parameters illustrated above, the stochastic fatigue truck model can be established via a Monte Carlo simulation. A linear growth factor

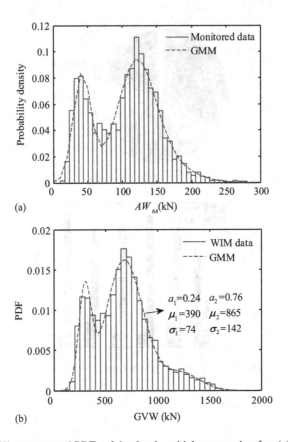

(a)

(b)

FIGURE 6.2 Histograms and PDFs of the 6-axle vehicle accounting for: (a) GVW; and (b) axle weight.

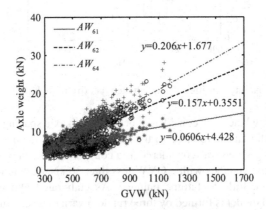

FIGURE 6.3 Fitted functions between axle weights and GVWs of the 6-axle truck.

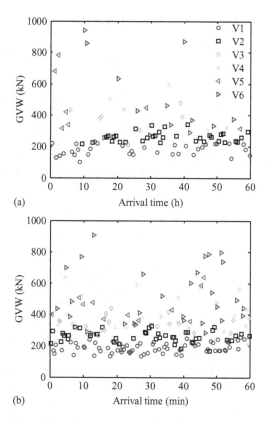

FIGURE 6.4   Simulated stochastic truck load models considering: (a) current traffic; and (b) future traffic.

of 0.5% was assumed for average daily traffic volume (ADTT). The simulated fatigue truck load model for 60 min for the current and the 100th year is shown in Figure 6.4. Note that the individual GVW is used instead of the axle weight for the convenience of expression.

As shown in Figure 6.4, each symbol indicates a truck specified by a different mark style, x-axis shows the arrival time, and y-axis shows the individual GVW. It is observed that although each individual truck differs to another, they follow a relative probability distribution. Therefore, the stochastic truck load model contains the statistics of the WIM measurements.

6.3   LIMIT STATE FUNCTION OF FATIGUE DAMAGE ACCUMULATION

On the basis of the fatigue damage accumulation formulations, a limit state function (LSF) of the fatigue damage accumulation was established with consideration of the traffic growth factors.

6.3.1 Fatigue Damage Accumulation Formulations

Each truck's passage will induce fatigue stress blocks at the welded joint of the bridge deck, and the fatigue stress blocks are both time and amplitude variant. Therefore, a calculation of the truck-induced fatigue damage accumulation should satisfy the requirements below. First, an S-N curve should be determined by considering the low-stress and high-cycle properties of truck-induced fatigue stress blocks. Second, categories of the welded joints should be included in the S-N curve. In the present study, the Eurocode3 specification (European Committee for Standardization (ECS), 2005) was used because of the thorough consideration of the properties illustrated above. The general form of the S-N curve in Eurocode3 specification is

$$\Delta\sigma_R^3 N_R = K_C \quad \left(\Delta\sigma_R \geq \Delta\sigma_D\right) \tag{6.1}$$

$$\Delta\sigma_R^5 N_R = K_D \quad \left(\Delta\sigma_L < \Delta\sigma_R \leq \Delta\sigma_D\right) \tag{6.2}$$

where, $\Delta\sigma_R$ is a fatigue stress range, N_R is the number of stress cycles, $\Delta\sigma_D$ and $\Delta\sigma_L$ are the constraint-amplitude fatigue threshold and the variant-amplitude fatigue threshold, respectively, and K_C and K_D are the detail coefficients of stress ranges that are higher and lower than $\Delta\sigma_D$, respectively. Taking the welded details of rib-to-deck and butt joint of U-rib as examples, the parameters of the S-N curve in Eurocode3 specification are shown in Table 6.2.

It is worth noting that the S-N curves can only be used for constant-amplitude fatigue blocks. However, the truck-induced fatigue stress ranges are variant-amplitude due to the randomness of the traffic loading. Thus, an equivalent stress range formula was utilized in the present study based on Miner's fatigue damage accumulation theory (Miner, 1945), which is written as

$$D = \sum_{\Delta\sigma_i \geq \sigma_D} \frac{n_i \Delta\sigma_i^3}{K_C} + \sum_{\Delta\sigma_j < \sigma_D} \frac{n_j \Delta\sigma_j^5}{K_D} = \frac{N_{eq} \Delta\sigma_{re}^5}{K_D} \tag{6.3}$$

where, D is the fatigue damage accumulation, $\Delta\sigma_i$ and $\Delta\sigma j$ are fatigue stress ranges that are greater and lesser than $\Delta\sigma_D$, respectively; n_i and n_j are the number of stress cycles for $\Delta\sigma_i$ and $\Delta\sigma_j$, respectively; $\Delta\sigma_{re}$ and N_{eq} are the equivalent fatigue stress range and the equivalent number of stress cycles, respectively. Note that N_{eq} under an individual truck load is assumed to be equal to the number of the axles. According to Equation (6.3), the $\Delta\sigma_{re}$ under an individual truck load can be written as

---

TABLE 6.2
S-N Curves Specified in the Eurocode3 Code

Welded joint	Detail category (MPa)	$\Delta\sigma_D$ (MPa)	$\Delta\sigma_L$ (MPa)	K_C	K_D
Deck-to-rib joint	50	37	20	2.50×10^{11}	7.16×10^{14}
Butt joint of U-ribs	71	52	29	3.47×10^{11}	19.00×10^{14}

$$\Delta\sigma_{re}^5 = \frac{\displaystyle\sum_{\Delta\sigma_i \geq \Delta\sigma_D} \frac{n_i\Delta\sigma_i^3}{K_C} + \sum_{\Delta\sigma_j < \Delta\sigma_D} \frac{n_j\Delta\sigma_j^5}{K_D}}{N_{eq}/K_D} \tag{6.4}$$

6.3.2 Limit State Function

The stress spectrum of daily traffic flow can be formed by conducting each truck passage analysis. In addition to the fatigue stress spectrum, the transverse distribution factor of the truck on the bridge deck and the traffic volume will affect the fatigue damage accumulation. During the long-term service period of a bridge, the ADTT and individual GVW will increase due to the development of the global economy. Considering all of the above parameters, the LSF of fatigue damage accumulation is written as

$$g_n(X) = D_\Delta - \sum_{t=1}^{n} D_t(X) = D_\Delta - 365N_d\Delta\sigma_{re}^5 w$$

$$\times \sum_{t=1}^{n} \left[1+(t-1)R_{ADTT}\right]\left[1+(t-1)R_{GVW}\right]^5 /K_D \tag{6.5}$$

$$N_d = N_{ADTT} \sum_{i=1}^{6} P_{(i)}N_{eq(i)} \tag{6.6}$$

where, D_Δ is the critical fatigue damage, D_t is the fatigue damage accumulation in the tth year, n (in years) is the service period of a bridge, w is the transverse distribution factor of a truck diving in a traffic lane, N_d is the number of daily stress cycles, $P_{(i)}$ is the occupancy rate of the ith vehicle type, N_{ADTT} is the number of ADTT, and R_{ADTT} and R_{GVW} are linear annual growth rates of the ADTT and GVW, respectively.

6.4 PROPOSED COMPUTATIONAL FRAMEWORK

In general, the traditional approach for calculating the truck-induced fatigue damage accumulation includes three steps (Chen et al. 2011; Zhang and Cai, 2011; Wang et al. 2013). First, simulate the stress history of a fatigue-critical point under a moving truck load. Second, convert the history into stress blocks using the rain-flow method. Finally, the individual fatigue damage is evaluated by the S-N curve and accumulated on the basis of Miner's rule. However, this approach is inappropriate for the stochastic truck load, because of the time-consuming problem caused by numerous repeating computer runs. Therefore, an efficient computational framework is presented for utilizing the stochastic truckload model for probabilistic modeling of the fatigue stress range.

Since the time-consuming problem origins from a large number of stress histories by repeating finite element analysis, a meta-model can be utilized to substitute the FE

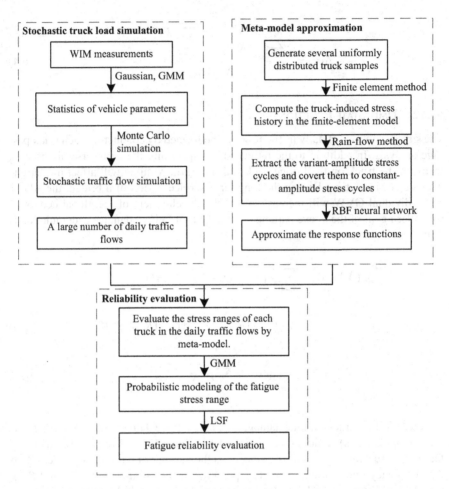

FIGURE 6.5 Flowchart of the proposed computational framework.

model. In the present study, a meta-model approximated by neural networks was used to substitute the time-consuming finite element simulation procedures. The flowchart of the procedures is summarized in Figure 6.5. The critical steps in the flowchart include the neural network-based response function approximation and GMM-based probabilistic modeling of the stress range.

The first critical step is to approximate the response surface functions. Since each truck passage will pose several stress blocks due to the multi-axle and dynamic effects, the response function between the axle load and the equivalent stress ranges is complex. An integration of the uniform design (UD) approach (Cheng, 2010; Liu et al. 2014) and radial basis function (RBF) neural networks shown in Figure 6.6, which are commonly used for structural reliability evaluation (Cheng et al. 2008; Liu et al. 2016), were utilized herein as a meta-model to approximate the implicit

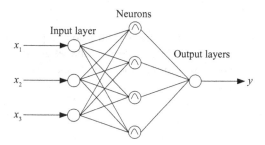

FIGURE 6.6 Diagram of an RBF neural network.

function between the axle weights and the equivalent stress ranges. In Figure 6.6, the input parameters in the present study indicate the axle weights for a vehicle type. Since there are six types of vehicle configuration, a total number of six meta-models were necessary for all the trucks. First of all, the upper and lower bound of the GVW should be determined, and then several uniformly distributed samples in the defined region generated via a UD approach. Subsequently, the FE analysis to compute the stress histories under individual truck passage is conducted and then the stress histories converted into stress blocks via the rain-flow method. Finally, the variant-amplitude stress blocks were converted to constant-amplitude stress blocks to be used to approximate the response function. The influence of the number of training samples on the prediction of the stress range will be discussed in the case study.

The second critical step is the probabilistic modeling based on GMM. With the approximated response function, the probabilistic modeling can be carried out efficiently. The purpose of the latter is to establish a probability model of an equivalent stress range. However, impacted by the probability distribution of random variable in the stochastic traffic flow, the PDF of the equivalent stress range is complex. Thus, the PDF of the stress range may not be a good fit with a single Gaussian or lognormal distribution function. In order to deal with this problem, a GMM was used to capture the characters of the bimodal or trimodal distribution of the PDFs. The GMM is a part of the finite mixture distributions which are commonly used for modeling complex probability distributions and which enable the statistical modeling of random variables with multimodal behavior. Considering Gaussian function as the given predictive component density, the GMM is written as (Xia et al. 2012).

$$f(y,a,\theta) = \sum a_i f_i(y|\theta_i) = \sum a_i \frac{1}{\sqrt{2\pi}} \exp\left\{-\frac{1}{2}\frac{(y-\mu_i)}{\sigma_i^2}\right\} \tag{6.7}$$

where, $f(y, a, \theta)$ is a predictive mixture density, $f(y|\theta_i)$ is a given parametric family of predictive component densities, a_i is the ith component weight, θ_i is the component parameters, and μ_i and σ_i are the mean value and standard deviation of the ith mixture parameter, respectively. The GMM provides an effective connection between the PDFs of GVWs and the equivalent stress range.

6.5   CASE STUDY

A steel box-girder of a suspension bridge was utilized as a prototype to demonstrate the application of the computational framework for the fatigue reliability assessment. The influences of the parameters in the stochastic truck load model on the fatigue reliability index are discussed.

6.5.1   BRIDGE DETAILS

Nanxi Bridge is a long-span highway suspension bridge in Sichuan, China. The orthotropic deck of the steel box-girder directly supports vehicular loads and contributes to the bridge's structural overall load-bearing behavior. Dimensions of the bridge are shown in Figure 6.7. The WIM measurements in Table 6.1 were collected from the SHM system (Liu et al. 2015) of this bridge.

An FE model of this bridge was simulated by the commercial FE program ANSYS as shown in Figure 6.8. In the global model, the beams and pylons were simulated by Beam188 elements, and the cables and hangers were simulated by Link180 elements. In the local model, all of the components were simulated by Shell63 elements. The longitudinal, transverse, and vertical length of the local model is 12.8 m, 14 m and 3 m respectively. The deck and U-ribs were meshed by quadrilateral elements, while the longitudinal stiffening plates, the diaphragm plates, and the web plates were meshed by triangular elements.

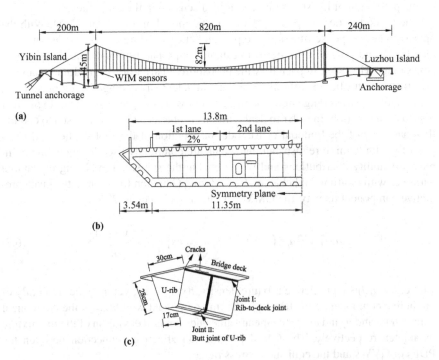

FIGURE 6.7   Dimensions of the bridge: (a) overall view; (b) a half cross-section; and (c) U-rib.

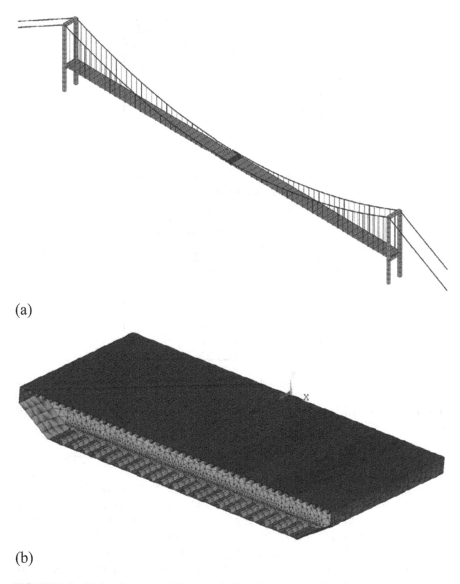

(a)

(b)

FIGURE 6.8 Finite element models of the bridge: (a) global model; (b) local model.

6.5.2 FINITE ELEMENT SIMULATION

The local model is in the mid-span point of the global model. In order to observe the truck-induced stress history, stress influence lines of the two welded joints in the global model were computed. The effective stress history is in the range of a segmental girder. Therefore, the static stress influence lines of the welded joints in the region of a segmental model are plotted in Figure 6.9. It can be seen that the effective stress influence line is mostly confined to the region of two diaphragm plates. This

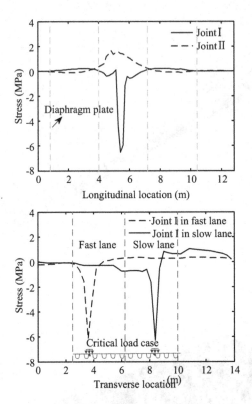

FIGURE 6.9 Stress influence lines of welded joints: (a) longitudinal direction; (b) transverse direction.

demonstrates the significance of the axle spacing and the vehicle configuration in the stochastic truck load model.

Note that the pavement has not been considered in the FE model. The axle load was simulated by a vertical uniformly distributed load that extends to the bridge deck at 45° (Guo et al. 2008). For instance, if the thickness of pavement is 6.7 cm, and the load area of the back wheel is 60 cm × 20 cm, then the revised load area is 73.4 cm × 33.4 cm. Under the truck load of the 6-axle trucks with maximum and minimum GVW, the stress histories of the rib-to-deck joint were computed as shown in Figure 6.10.

As observed in Figure 6.10, there are six peaks corresponding to the six axles of the truck. Each variant-amplitude stress cycle is posed by an axle load. For a 2-axle truck, 20 training data were designed as uniformly distributed samples. Subsequently, 20 times of computer runs were conducted to evaluate the equivalent stress ranges corresponding to the design samples. The axle weight and the equivalent stress range are input and output data of the response function. A MATLAB toolbox entitled NEWRBF (neural networks of radial basis function) was used to train the RBF neural network, where the number of neurons equals the number of input samples, and the kernel function was the Gaussian function. The response function of the approximated neural network is shown in Figure 6.11.

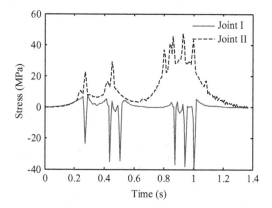

FIGURE 6.10 Stress-time histories of the rib-to-deck joint under a standard 6-axle truck load.

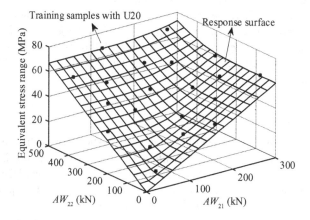

FIGURE 6.11 Response surface of the stress range under the 2-axle truck load.

As shown in Figure 6.11, the design samples are uniformly distributed in the coordinate of variables, and the approximated function is nonlinear and appropriate. With the approximated response function of each type of vehicle, the fatigue stress ranges of the rib-to-deck joint under the daily traffic flow were calculated. It is acknowledged that the accuracy of the meta-model depends on the number of training samples, which will also therefore determine the computational effort. Therefore, the influence of the number of training samples on the accuracy of the meta-model was analyzed. The accuracy of the meta-model was reflected by the root-mean-square error (RMSE) of 100 random samples following uniform distribution. The RMSE of the approximated neural network of the 2-axle truck (V2) and the 6-axle truck (V6) is shown in Figure 6.12.

As can be seen in Figure 6.12, the 6-axle truck needs more training samples to ensure the accuracy of the meta-model. This model includes three independent axle load variables: the front axle load, the tandem axle, and the tridem axle. Under the

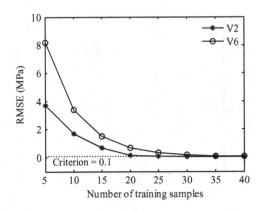

FIGURE 6.12   Influence of the number of training samples on the RMSE of stress ranges.

accuracy criterion of 0.1 MPa, the 2-axle and 6-axle truck model need 20 and 40 samples, respectively. Therefore, the six types of truck load model need 180 training samples. The entire finite element computational effort for a core-7 computer is roughly two hours. However, without the meta-model, one-day traffic of an average 2000 trucks would need 22 hours. The computational effort for a 100-day traffic analysis is inconceivable without the meta-model.

6.5.3   PROBABILISTIC MODELING

The PDF of the stress range under the 100-day stochastic truck loading fits the GMM, and the daily number of stress cycles fits the Gaussian distribution. Note that the number of stress cycles equals the total number of vehicle axles. Histograms and PDFs of the stress range and the number of cycles of the rib-to-deck joint under the stochastic fatigue truck load in 100 days are shown in Figure 6.13.

It can be found from Figure 6.13 that there are three peaks in the probability density of the stress range that is well fitted by the GMM compared to the normal distribution model. As elaborated above, the probability model of the stress range and the number of cycles established above provide a basis for the subsequent fatigue reliability evaluation.

In addition to stress spectrum, the fatigue strength of welded joints is also critical for fatigue reliability evaluation. This study assumed the fatigue strength coefficient in terms of resistance following lognormal distributions. The transverse distribution factor of the truck axle is assumed to follow normal distribution with the mean value of 0.3 and coefficient of variation (COV) of 1. Based on the above assumption, the statistics of the variables in the LSF are shown in Table 6.3.

6.5.4   FATIGUE RELIABILITY EVALUATION

With the probability model of the fatigue stress range, the fatigue reliability can be evaluated in consideration of the service period of the bridge. In practice, the traffic volume and GVW will increase with the development of a society's economy.

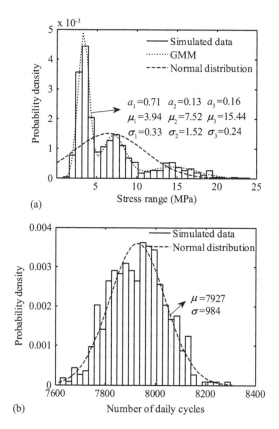

$a_1=0.71$ $a_2=0.13$ $a_3=0.16$
$\mu_1=3.94$ $\mu_2=7.52$ $\mu_3=15.44$
$\sigma_1=0.33$ $\sigma_2=1.52$ $\sigma_3=0.24$

$\mu=7927$
$\sigma=984$

FIGURE 6.13 Histograms and PDFs of the rib-to-deck joint under the stochastic truck load: (a) stress range, and (b) number of daily cycles.

Suppose the annual linear growth rates of the GVW and ADTT are constants to a range of 0–0.5%. On this basis, the fatigue reliability indices are shown in Figure 6.14.

It is observed from Figure 6.14 that the growth rate of both the ADTT and the GVW lead to a rapid decrease in the reliability index. A growth rate of ADTT results in a significant decrease of the fatigue reliability from 5.94 to 2.87. However, a growth rate of the GVW results in a more significant decrease from 5.94 to 0.92. This has demonstrated that the growth rate of GVW results in a faster decrease of the reliability index compared to that caused by the growth of ADTT. In practice, this phenomenon can be explained by the LSF shown in Equation (6.5). Therefore, the control of overloaded trucks rather than the traffic volume is an effective way to ensure the fatigue safety of orthotropic steel bridge decks.

Since overloading is the major reason for the fatigue failure of orthotropic steel bridge decks, influence of the overload rate on the fatigue reliability is worthy of investigation. According to the Chinese specification (MOCAT, 2004), the threshold of GVW for 2-axle and 6-axle trucks is 200 kN and 550 kN, respectively. In the present study, the threshold overload rate was assumed to be 25%, 50%, 75% and 100%. Based on the above assumption, the stochastic truck load model was updated with the

TABLE 6.3
Statistics of Random Variables

Random variable	Symbol	Mean value	COV	Distribution
Critical fatigue damage	D_Δ	1.0	0.3	Lognormal
Fatigue strength coefficient	K_D	3.47×10^{14}	0.34	Lognormal
Transverse distribution factor	w	0.8	1	Normal
Equivalent stress range	$\Delta\sigma_{re}$	See Figure 6.13(a)		GMM
Number of daily cycles	N_d	See Figure 6.13(b)		Gaussian

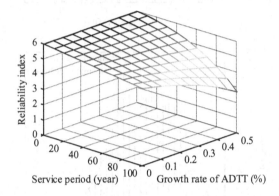

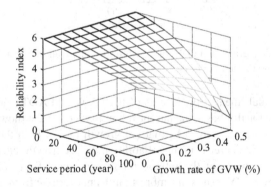

FIGURE 6.14 Fatigue reliability index of the rib-to-deck joint considering traffic growth in: (a) ADTT; (b) GVW.

predefined threshold overload rate. The influence of the threshold overload rate on the fatigue reliability of the rib-to-deck joint in the 100th year is shown in Figure 6.15.

It is observed from Figure 6.15 that the threshold overload rate has a significant impact on the fatigue reliability index. Even for a threshold of 100%, the fatigue reliability index has an obvious increase. However, this increase trend weakens with stricter controls on the overload rate. This result suggests that even though there has been a rapid growth in traffic volume, the control of overloaded trucks is an effective way to ensure the fatigue reliability of steel bridges.

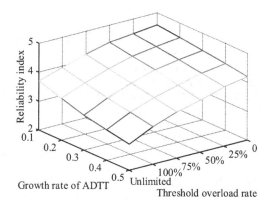

FIGURE 6.15   Influence of the threshold overload rate on the fatigue reliably index.

6.6 CONCLUSIONS

A stochastic truck load model was developed based on site-specific WIM measurements for fatigue reliability evaluation of orthotropic steel bridge decks. The time-consuming problem of the finite element-based fatigue stress simulation was solved by utilizing a meta-model approximated by neural networks. The effectiveness of the stochastic truck load model in probabilistic modeling and fatigue reliability assessment was demonstrated by the case study of a steel box-girder bridge. The following conclusions are drawn:

1. The computational effort of the fatigue stress analysis in the FE model is greatly reduced by utilizing the meta-model. However, the accuracy of the prediction mostly depends on the number of the training samples with a uniform design scheme. Approximately 180 training samples is sufficient for the meta-models of the six types of vehicles.
2. Compared to the traffic volume increase, the growth of GVWs has a more significant impact on the bridge fatigue reliability. A growth rate of ADTT results in a significant decrease of the fatigue reliability from 5.94 to 2.87. However, a growth rate of the GVW results in a more significant decrease from 5.94 to 0.92.
3. The threshold overload rate reduces the decreased range of reliability index caused by the ADTT growth. Even though the traffic volume is growing rapidly, the control of extremely overloaded trucks in comparison to the traffic volume is an effective way to ensure the fatigue safety of orthotropic steel bridge decks.

Future efforts are needed to improve the stochastic fatigue truck load model by considering the vehicle spacing parameter. The neural network approach can be replaced by an advanced approach to make the computational framework more efficient. In addition, the vehicle–bridge interaction and the degeneration of road surface roughness condition will be considered in future work.

REFERENCES

Biezma, M. V. and Schanack, F. 2007. Collapse of steel bridges. *Journal of Performance of Constructed Facilities* 21(5): 398–405.

Chen, S. R. and Wu, J. 2011. Modeling stochastic live load for long-span bridge based on microscopic traffic flow simulation. *Computers and Structures* 89(9): 813–824.

Chen, Z. W., Xu, Y. L., Xia, Y., Li, Q. and Wong, K. Y. 2011. Fatigue analysis of long-span suspension bridges under multiple loading: case study. *Engineering Structures* 33(12): 3246–3256.

Cheng, J. 2010. An artificial neural network based genetic algorithm for estimating the reliability of long span suspension bridges. *Finite Elements in Analysis and Design* 46(8): 658–667.

Cheng, J., Li, Q. and Xiao, R. 2008. A new artificial neural network-based response surface method for structural reliability analysis. *Probabilistic Engineering Mechanics* 23(1): 51–63.

Deng, Y., Liu, Y. Feng, D. M. and Li, A. Q. 2015. Investigation of fatigue performance of welded details in long-span steel bridges using long-term monitoring strain data. *Structural Control and Health Monitoring* 22(11): 1343–1358.

Dung, C. V., Sasaki, E. Tajima, K. and Suzuki, T. 2015. Investigations on the effect of weld penetration on fatigue strength of rib-to-deck welded joints in orthotropic steel decks. *International Journal of Steel Structures* 15(2): 299–310.

European Committee for Standardization (ECS). (2005). *Eurocode 3: Design of steel structures—Part 1–9: Fatigue EN1993-1-9*. Brussels, Belgium.

Frangopol, D. M., Strauss, A. and Kim, S. 2008. Bridge reliability assessment based on monitoring. *Journal of Bridge Engineering* 13(3): 258–270.

Gokanakonda, S., Ghantasala, M. K. and Kujawski, D. 2016. Fatigue sensor for structural health monitoring: Design, fabrication and experimental testing of a prototype sensor. *Structural Control and Health Monitoring* 23(2): 237–251.

Guo, T., Li, A. and Wang, H. 2008. Influence of ambient temperature on the fatigue damage of welded bridge decks. *International Journal of Fatigue* 30(6): 1092–1102.

Guo, T., Frangopol, D. M. and Chen, Y. 2012. Fatigue reliability assessment of steel bridge details integrating weigh-in-motion data and probabilistic finite element analysis. *Computers and Structures* 112: 245–257.

Guo, T. and Chen, Y. W. 2013. Fatigue reliability analysis of steel bridge details based on field-monitored data and linear elastic fracture mechanics. *Structure and Infrastructure Engineering* 9(5): 496–505.

Ji, B., Liu, R. Chen, C. Maeno, H. and Chen, X. 2013. Evaluation on root-deck fatigue of orthotropic steel bridge deck. *Journal of Constructional Steel Research* 90: 174–183.

Lalthlamuana, R. and Talukdar, S. 2013. Rating of steel bridges considering fatigue and corrosion. *Structural Engineering and Mechanics* 47(5): 643–660.

Liu, Y., Lu, N. Noori, M. and Yin, X. 2014. System reliability-based optimization for truss structures using genetic algorithm and neural network. *International Journal of Reliability and Safety* 8(1): 51–69.

Liu, Y., Lu, N. W. and Yin, X. W. 2016. A hybrid method for structural system reliability-based design optimization and its application to trusses. *Quality and Reliability Engineering International* 32(2): 595–608.

Liu, Y., Deng, Y. and Cai, C. S. 2015. Deflection monitoring and assessment for a suspension bridge using a connected pipe system: a case study in China. *Structural Control and Health Monitoring* 22(12): 1408–1425.

Lydon, M., Taylor, S. E. Robinson, D. Mufti, A. and Brien, E. J. O. 2016. Recent developments in bridge weigh in motion (B-WIM). *Journal of Civil Structural Health Monitoring* 6(1): 69–81.

Marques, F., Moutinho, C. Hu, W. H. Cunha, Á. and Caetano, E. 2016. Weigh-in-motion implementation in an old metallic railway bridge. *Engineering Structures* 123: 15–29.

Miner, M. 1945. Cumulative damage in fatigue. *Journal of Applied Mechanics*. 12(3): 159–164.

Ministry of Communications and Transportation (MOCAT). (2004). *Limits of dimensions, axle load and masses for road vehicles GB 1589–2004*, China Communications Press, Beijing, China.

Ni, Y. Q., Ye, X. W. and Ko, J. M. 2010. Monitoring-based fatigue reliability assessment of steel bridges: analytical model and application. *Journal of Structural Engineering* 136(12): 1563–1573.

OBrien, E. J. and Enright, B. 2013. Using weigh-in-motion data to determine aggressiveness of traffic for bridge loading. *Journal of Bridge Engineering* 18(3): 232–239.

OBrien, E. J., Cantero, D. Enright, B. and González, A. 2010. Characteristic dynamic increment for extreme traffic loading events on short and medium span highway bridges. *Engineering Structures* 32(12): 3827–3835.

Sim, H. B. and Uang, C. M. 2012. Stress analyses and parametric study on full-scale fatigue tests of rib-to-deck welded joints in steel orthotropic decks. *Journal of Bridge Engineering* 17(5): 765–773.

Sim, H. B., Uang, C. M. and Sikorsky, C. 2009. Effects of fabrication procedures on fatigue resistance of welded joints in steel orthotropic decks. *Journal of Bridge Engineering* 14(5): 366–373.

Wang, T. L., Liu, C., Huang, D. and Dahawy, M. 2005. Truck loading and fatigue damage analysis for girder bridges based on weigh-in-motion data. *Journal of Bridge Engineering* 10(1): 12.

Wang, D., Zhang, D. Wang, S. and Ge, S. 2013. Finite element analysis of hoisting rope and fretting wear evolution and fatigue life estimation of steel wires. *Engineering Failure Analysis* 27: 173–193.

Xia, H. W., Ni, Y. Q., Wong, K. Y. and Ko, J. M. 2012. Reliability-based condition assessment of in-service bridges using mixture distribution models. *Computers and Structures* 106: 204–213.

Ye, X. W., Ni, Y. Q.. Wong, K. Y. and Ko, J. M. 2012. Statistical analysis of stress spectra for fatigue life assessment of steel bridges with structural health monitoring data. *Engineering Structures* 45: 166–176.

Ye, X. W., Yi, T. H. Wen, C. and Su, Y. H. 2015. Reliability-based assessment of steel bridge deck using a mesh-insensitive structural stress method." *Smart Structures and Systems* 16(2): 367–382.

Zhang, J. and Au, F. 2016. Fatigue reliability assessment considering traffic flow variation based on weigh-in-motion data. *Advances in Structural Engineering*. 20(1), 125–138.

Zhang, W. and Cai, C. S.. 2011. Fatigue reliability assessment for existing bridges considering vehicle speed and road surface conditions. *Journal of Bridge Engineering* 17(3): 443–453.

Zhao, J. and Tabatabai, H. 2012. Evaluation of a permit vehicle model using weigh-in-motion truck records. *Journal of Bridge Engineering* 17(2): 389–392.

Zhou, Y. and Chen, S. 2015. Dynamic Simulation of a long-span bridge-traffic system subjected to combined service and extreme loads. *Journal of Structural Engineering* 141(9): 04014215.

7 Probabilistic Fatigue Damage of Orthotropic Steel Deck Details based on Structural Health Monitoring Data

Haiping Zhang

Hunan University of Technology, China

Yang Liu

Hunan University of Technology, China

Naiwei Lu

Changsha University of Science and Technology, China

CONTENTS

7.1   INTRODUCTION

The orthotropic steel deck has been widely used in the girder structure of long-span bridges. It suffers from fatigue damage because of the high fatigue stress from the vehicle and environment loads. For this reason, a more detailed fatigue reliability assessment for steel bridges was needed.

Fatigue reliability assessments for orthotropic steel deck bridge have been well developed. The methods used for fatigue reliability analysis can be divided into fatigue strain statistical analysis according to strain sensor data (Deng and Ding 2011) and finite element analysis (FEA) based on weigh-in-motion (WIM) data (Guo et al. 2012).The FEA method can calculate all the focus points of stress. However, these approaches simply cannot consider the influence of both the finite element model and ambient loadings. The fatigue reliability analysis method became more available with the development of structural health monitoring. A typical fatigue limit state function includes two types of random variables: fatigue strength and fatigue load spectrum (Kwon and Dan, 2010).

The values of the parameters were provided by AASHTO specifications (2011). The fatigue effect variables include two main aspects: fatigue stress range and fatigue loading cycles. Typically, researchers have treated fatigue stress range and fatigue loading cycles as two independent variables (Ni et al. 2010; Liu et al. 2010). However, there may be correlation between the two variables due to the regional characteristic of traffic flows (Liu et al. 2016). Thus, the values of fatigue reliability index are not accurate unless the correlation of these variables is considered. For this reason, the joint function of fatigue effect variables should be developed.

Recently, the copula function has been widely used in the fields of finance (Fengler and Okhrin, 2014) and hydrology (Singh and Zhang 2006) because of its powerful connect function and concise expression. The copula function offers an agile way for modeling the joint function based on multivariate data (Tang et al. 2015; Eryilmaz et al. 2014; Tang et al. 2013a). For example, Wu (2014) modeled the joint probability function using a copula approach. However, few studies have used copula in fatigue reliability assessment for the following reasons: (1) Lack of in-depth study for SHM data mining. The correlation of variables was ignored by researchers because of a lack of complete data information from health monitoring systems. (2) The process of fatigue reliability index calculation becomes complicated when using copula. It is difficult to obtain the expression of the display joint probability function.

In this chapter, the approach of fatigue reliability assessment based on SHM data using copula is proposed. The main contribution of this investigation is twofold. First, a system of fatigue reliability assessment using copula was established and the types of correlation of the monitoring data analyzed. Different types of copula functions were explored to find the optimal fitting function. Second, a method of fatigue reliability index calculation for joint probability distribution was proposed.

7.2   FATIGUE RELIABILITY ASSESSMENT INTEGRATING SHM DATA USING COPULA

7.2.1 Fatigue Reliability Functions Based on SHM Data

The design of a bridge structure demands that its load effect S is lower than the resistance R, so the limit station equation can be simply expressed as follows:

$$g(X) = R - S = 0 \tag{7.1}$$

where $X = \{X_1, X_2, \ldots, X_n\}$ denotes the random variables.

The AASHTO design specification and the Miner rule (Miner 1945) gives the information associated with R for bridge fatigue assessment. The fatigue stress spectrum which provides the load effect S calculated by the rain-flow counting method from the SHM data, according to Equation 7.1, can be written into

$$g(X) = \frac{A \cdot \Delta}{e \cdot S_{eq}^B} - N_m = 0 \tag{7.2}$$

where A is the fatigue strength coefficient, B is the fatigue strength exponent, which has a value of 3.0, Δ is the Miner's critical damage accumulation index which can be assigned as 1.0, and e is the measurement error in SHM. N_t is the numbers of equivalent constant stress cycles form the start of fatigue damage to the time m. S_{eq} is the equivalent constant stress range, which can be calculated as follows:

$$S_{eq} = \left[\sum \frac{n_i}{N_t} S_i^B \right]^{\frac{1}{B}} \tag{7.3}$$

$$N_t = 365 \cdot m \cdot N_d \cdot \int_0^{T_m} (1+\alpha)^m \, dm \tag{7.4}$$

where n_i is the number of stress cycles corresponding to the ith stress range, S_i. N_d is the daily number of stress cycles corresponding to the equivalent stress S_{eq}, m is the value of service years for the bridge, and α is the annual growth rate of traffic.

According to Equations (7.2), (7.3) and (7.4), the function for fatigue reliability assessment based on SHM data was established as follows:

$$g(X) = \Delta - e \cdot \frac{365 \cdot m \cdot \int_0^{T_m} (1+\alpha)^m \, dm \cdot N_d \cdot S_{eq}^B}{A} \tag{7.5}$$

Table 7.1 shows the parameters in Equation (7.5).

TABLE 7.1
The Parameters of Fatigue Reliability Limit Equation

Parameters	Distribution type	source
Damage accumulation index, Δ	Lognormal, LN (1.0, 0.3)	Wirsching (1984)
Fatigue coefficient, A (MPa³)	Normal, N (1.44e12, 6.48e12)	Zhao et al. (1994)
Error factor of measurement, e	Lognormal, LN (1.0, 0.04)	Dan et al. (2008)
Design of bridge life, T_d	Deterministic, 100a	Bridge data
Annual increase rate, α	Deterministic, (0%,2% 5%)	Guo and Chen (2011)

7.2.2 Reliability Index

The expression of fatigue failure probability as follows:

$$p_f = \iint_{g(X)<0} \cdots \int g_X(x_1,x_2,\cdots,x_n) dx_1 dx_2 \cdots dx_n \qquad (7.6)$$

where $g_{xn}(x_1, x_2,\ldots,x_n)$ is the joint probability function of $g_{x1}(x_1)$, $g_{x2}(x_2)$, $\ldots,g_{xn}(x_n)$.

The expression of reliability index can be expressed as

$$\beta = \Phi^{-1}(1 - P_f) \qquad (7.7)$$

where Φ^{-1} is the inverse standard normal distribution

7.2.3 Theory of Copula

As Equation (7.6) shows, in order to calculate the fatigue failure reliability, the joint density function needs to be obtained. However, we can only establish every variable's marginal density distribution based on the incomplete measured data. Consider a joint function $G(x_1, x_2)$ which is associated with marginal function $G_1(x_1)$ and $G_2(x_1)$ (Zhao et al. 1994).

$$G(x_1,x_2) = C\big(G_1(x_1),G_2(x_2),\theta\big) \qquad (7.8)$$

where $C(\cdot)$ is a type of copula. θ is the parameter describing the correlation between random variables X_1 and X_2. The value of θ determined by the line correlation coefficient, ρ, can be written as follows (Ang and Tang, 2007):

$$\rho = \frac{\text{cov}(x_1,x_2)}{\sigma_1\sigma_2} = \int_{-\infty}^{+\infty}\int_{-\infty}^{+\infty}\left(\frac{x_1-\mu_1}{\sigma_1}\right)\left(\frac{x_2-\mu_2}{\sigma_2}\right)g_1$$
$$\times(x_1)g_2(x_2)c\big(G_1(x_1),G_1(x_1);\theta\big)dx_1dx_2 \qquad (7.9)$$

where $\text{cov}(x_1, x_2)$ is the covariance for random variables X_1 and X_2, μ_1 and σ_1 are the mean and standard deviation of X_1, μ_2 and σ_2 are the mean and standard deviation of X_2, $c(\cdot)$ is the probability density distribution (PDF) of $C(\cdot)$, $g_1(x_1)$, and $g_2(x_2)$ are PDFs of $G_1(x_1)$ and $G_2(x_2)$. The Sklar theory is also suitable for more variables. The expression can be shown as follows:

$$G(x_1,x_2,\ldots,x_n) = C\big(G_{x1}(x_1),G_{x2}(x_2),\ldots,G_{xn}(x_n),\theta\big) \qquad (7.10)$$

Taking a derivative of Equation (7.10), the joint probability function can be obtained:

$$g(x_1,x_2,\ldots,x_n) = c\big(G_{x1}(x_1),G_{x2}(x_2),\ldots,G_{xn}(x_n),\theta\big) \cdot \prod_i^n g_{xi}(x_i) \qquad (7.11)$$

where $g_i(x_i)$ is the PDF of $G_i(x_i)$. To simplify the expressions, use u_i to replace $G_{xi}(x_i)$. It is worth noting that the copula $C(\cdot)$ function has the following characteristics: (1) The domain of function $C(\cdot)$ is $u_i \in [0\ 1]$. (2) $C(\cdot)$ is monotonically increasing function corresponding to each random variable. (3) If the random variables $u_1, u_2, \ldots, u_n$ are independent, then $C(u_1, u_2, \ldots, u_n) = u_1 u_2 \ldots u_n$.

7.2.4 COMPARISON AND SELECTION OF COPULAS

There exist types of structural dependence between X_1 and X_2. Many copula functions, Gaussian, t-copula, Gumbel, Frank, and Clayton copulas, have been introduced to describe the correlation between variables. The Gaussian copula is widely used because of the simple expression and easy parameter estimation (Stefano and McNeil, 2005). The t-copula has a strong ability to capture extreme values. The Gumbel, Frank, and Clayton copulas belong to the Archimedean copulas, and the Gaussian copula and t-copula belong to the Gaussian copulas. The five types of expression for copulas (Akaike 1974) are listed in Table.7.2.

A criterion needs to be introduced to select the optimal copula function among the different types of copula listed in Table. 7.2. The Akaike information criterion (AIC) was adopted in this paper because of the simple explicit form. The AIC is defined as follows:

$$V_{AIC} = -2\ln(c) + 2a \qquad (7.12)$$

where $\ln(c)$ is the value of the sum of log-likelihood and a is the number of parameters. It is worth noting that the value of $\ln(c)$ is calculated by the maximum likelihood

TABLE 7.2
Types of Copulas and Expression

Types	CDF of copula, $C(u_1, u_2)$	PFD of copula, $c(u_1, u_2)$	Rang of θ
Gaussian copula	$\Phi_\theta(\Phi^{-1}(u_1), \Phi^{-1}(u_2))$	$\dfrac{1}{\sqrt{1-\theta^2}}\exp\left(-\dfrac{\xi_1^2\theta^2 - 2\xi_1\xi_2\theta + \xi_2^2}{2(1-\theta^2)}\right),$ $\xi_1 = \Phi^{-1}(u_1),\ \xi_2 = \Phi^{-1}(u_2)$	$[-1\ 1]$
t-copula	$t_\theta(t^{-1}(u_1), t^{-1}(u_2))$	$\dfrac{1}{2\pi\sqrt{1-\rho^2}}\left(1 + \dfrac{u_1^2 - 2\rho u_1 u_2 + u_2^2}{k(1-\rho^2)}\right)^{-(k+2)/2}$	$(0,1)$
Gumbel copula	$\exp\left(-\left[(-\ln u_1)^\theta + (-\ln u_2)^\theta\right]^{\frac{1}{\theta}}\right)$	-	$[1, \infty)$
Frank copula	$-\dfrac{1}{\theta}\ln\left(1 + \dfrac{(e^{-\theta u_1}-1)(e^{-\theta u_2}-1)}{e^{-\theta}-1}\right)$	$\dfrac{e^{-\theta u_1}(e^{-\theta u_2}-1)}{(e^{-\theta}-1)+(e^{-\theta u_1}-1)(e^{-\theta u_2}-1)}$	$(-\infty, \infty)$
Clayton copula	$\left(u_1^{-\theta} + u_2^{-\theta} - 1\right)^{-\frac{1}{\theta}}$	$u_1^{-\theta-1}\left(u_1^{-\theta} + u_2^{-\theta} - 1\right)^{-\frac{1}{\theta}-1}$	$(0, \infty)$

estimation method (Jiang et al. 2014). The best-fit copula is associated with the smallest value of V_{AIC}.

7.2.5 Fatigue Reliability Analysis Process Using Copula

Figure 7.1 shows the flowchart of the proposed computational framework. The process of fatigue reliability assessment using copula is divided into six main steps:

Step 1 Confirm the types of variables in the limit state function and the SHM data. There are two types of variables which need to be obtained: resistance variables (R) and load effect variables (S). For R, we can refer to the AASHTO specification. For S, the strain data from SHM needs to remove the interference using wavelet extraction technology.

Step 2 Analyze the correlation of variables. The values of variables need to be listed in time domain, and the variation laws for variables will be discussed.

Step 3 Select different types of copula functions to fit the variables. The correlation coefficient can be calculated by utilizing maximum likelihood method.

Step 4 The AIC was introduced to select the optimal copula function. The minimum value of AIC must correspond to the best copula function among those copulas.

Step 5 Establishing the joint function for fatigue reliability assessment on which the limit equation was built.

Step 6 The fatigue reliability index, β, was calculated with the Monte Carlo method.

7.3 CASE STUDY OF NANXI YANGTZE RIVER SUSPENSION BRIDGE

7.3.1 Health Monitoring System of Yangtze River Suspension Bridge

The main span of the Nanxi Yangtze River suspension bridge is 820 m, which is the longest bridge in southwest China (see Figure 7.2). To monitor bridge performance under complex loading, the health monitoring system was installed. Two types of monitoring data, structure strain and environment temperature data, were analyzed in this paper. The ambient temperature sensors measured the data at a frequency of 5 Hz. From the measured data over a period of 10 min, a mean temperature was calculated giving a total number of 52,560 average temperature measurements (Liu et al. 2017b). The sampling frequency of strain sensors was 10 Hz. Figure 7.3 shows the installation position of strain sensors.

7.3.2 Strain Data-processing

Strain data which were collected from sensors are influenced by vehicle loads, ambient temperature, and random disturbance. To analyze the fatigue strain cycles by vehicle loads, the strain caused by temperature and random disturbance need to be separated. There are vastly different frequencies for vehicle load, ambient

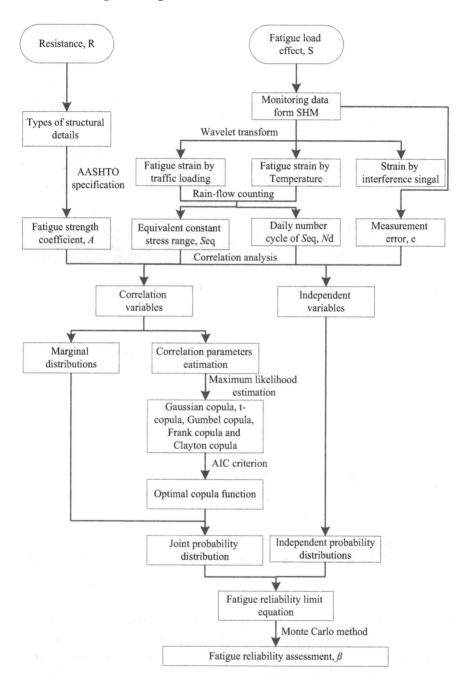

FIGURE 7.1   Flowchart of fatigue reliability assessment using copula based on SHM data.

FIGURE 7.2   Site photo of Nanxi Yangtze River suspension bridge.

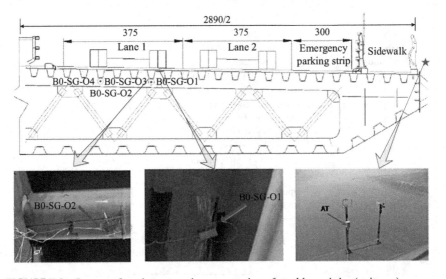

FIGURE 7.3   Layout of strain sensors in cross section of steel box girder (unit: cm).

temperature, and random disturbance. Wavelet Transfer (WT) is a powerful technique for signal separation (Keissar et al. 2010) as it is uncomplicated for WT to separate the random noise. A time-history curve of strain based on the measured data during May 1, 2015 can be divided into strain caused by temperature and that caused by traffic loads, as shown in Figure 7.4.

The highly sample frequency of strain sensors means a large number of useless data in the strain sample. The effective fatigue strains are usually associated with vehicle loads. (Downing and Socie 1982). However, the other strain points can disturb the strain curves. It is an effective way to extract the top and bottom points in the

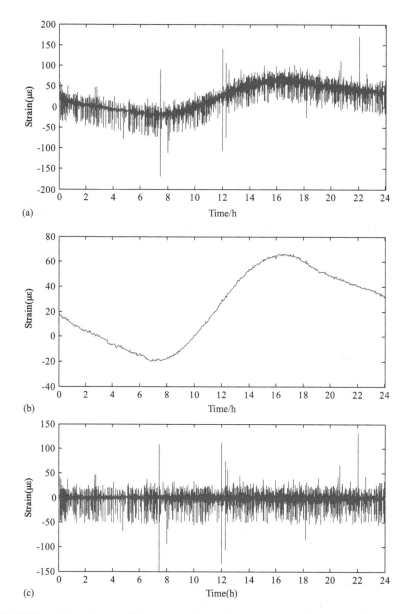

FIGURE 7.4   Time-history of the strain for BO-SG-01 during May 1, 2015: (a) the measured strain data; (b) the strain accounting for temperature; (c) the strain accounting for traffic loads.

strain curve, and the stress cycle is established by connecting the two points. The main procedures are explained considering three representative points including E_1, E_2, and E_3. If $E_1 < E_2 > E_3$ or $E_1 > E_2 < E_3$, E_2 is the top point or the bottom point, then E_2 will be selected and input into matrix A. Subsequently, select the next set of points and generate the new consecutive point E_1, E_2, $_{\text{and } E3.}$ In this way, all the key points for strain curve will be included in matrix A. For instance, select 100 samples in a strain

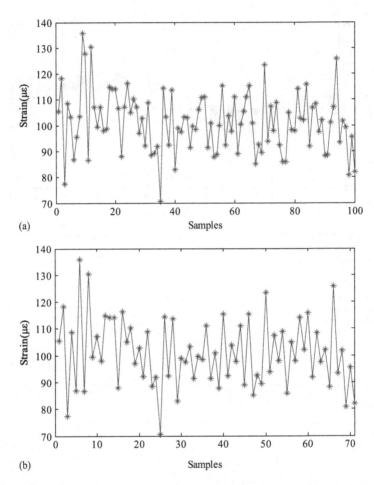

FIGURE 7.5 Strain inflection point extraction: (a) measured data; (b) inflection point.

sensor as an illustration example. Figure 7.5 shows that the efficiency of the strain processing procedures in compressing the measurements.

7.3.3 VARIABLES CORRELATION DISCUSSION

Critical fatigue damage Δ, fatigue strength coefficient A, error coefficient of strain sensors e, equivalent stress amplitude S_{eq} and daily stress cycles N_d were considered as random variables in Equation (7.7). According to Miner theory, the critical fatigue damage index equal to 1, which demonstrate the fatigue limit state.. A lognormal distribution with mean value of 1 and variable coefficient of 0.3 fits well (Nelsen, 2006). The value of A is associated by material categories. So, A is the independent variable in Equation (7.5). The measuring accuracy of strain sensors may be influenced by many factors because of the complex environment. Hence, the values of e are all dependent on the sensors and environment (Liu et al. 2016). It can be supposed that e has no relationship with other variables.

7.3.4 RELEVANT VARIABLES

If two variables are influenced by the same factor, there will be a relationship between the two variables. The fatigue stress effects have a linear relationship with ambient temperature. Meanwhile, fatigue stress cycles may be changed by temperature gradient effect and a change in the number of vehicles. There may be a relationship between S_{eq} and N_d. The daily extreme value of ambient temperature was obtained by temperature sensors. S_{eq} and N_d were calculated by Equation (7.3) based on collected data. The time-dependent curve for ambient temperature T, S_{eq}, and N_d is based on a contrastive analysis from Jan-01 to Jul-30 as shown in Figure 7.6. The daily fatigue equivalent stress and fatigue stress cycle numbers of U-rib-to-deck details (BO-SG-01) is sensitive to ambient temperature. S_{eq} and N_d of U-rib bottom (BO-SG-02) had little change from temperature.

7.3.4.1 Joint Function Modeling Using Copula

To establish the joint function, the optimal types of marginal distributions of u_1 and u_2 need to be selected. Normal, Lognormal, and Weibull models were used to describe the measured data of S_{eq} and N_d as shown in Figure 7.7 and Figure 7.8. The parameters of Normal, Lognormal, and Weibull models are listed in Table 7.3. Then, the Kolmogorov–Smirnov goodness of fit test was used as a criterion to select the optimal models. The result shows that both u_1 and u_2 accord with lognormal models' assumptions when confidence is equal to 95%.

The selection of optimal copula process is divided into two main steps. Step 1 calculates the correlation parameter. There are two main methods of calculation used to assess the correlation parameter; the Pearson method and the Kendall method. The Kendall method has more extensive calculation scope and simplified expression and was adopted to compute the correlation parameter. The relationship between Kendall correlation coefficient τ and correlation parameter θ can be simply expressed as follows:

$$\theta = \sin\left(\frac{\pi\tau}{2}\right) \qquad (7.13)$$

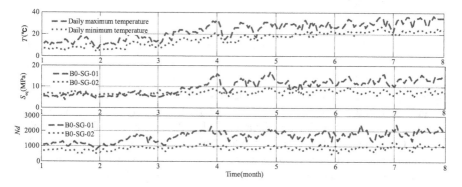

FIGURE 7.6   Comparisons for the time-history curves of T, S_{eq}, and N_d.

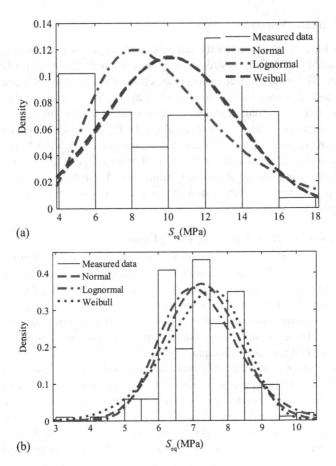

FIGURE 7.7 Measured data and fitted probability models of S_{eq}: (a) BO-SG-01; (b) BO-SG-02.

where τ can be expressed as

$$\tau = \binom{N}{2}^{-1} \sum_{i<j} sign\left[\left(x_{1i} - x_{1j}\right)\left(x_{2i} - x_{2j}\right)\right] \tag{7.14}$$

where x_{1i}, x_{1j}, x_{2i}, and x_{2j} is the observed data, N is the length of observed data, sign $(\cdot)$ is sign function, if $(x_{1i} - x_{1j})(x_{2i} - x_{2j}) > 0$, sign $(\cdot) = 1$ or else sign $(\cdot) = -1$. Step 2 is to calculate the value of AIC using Equation (7.12). The parameters for the t-copula, and the Gaussian, Gumbel, Frank, and Clayton copulas are listed in Table. 7.4. The minimum value of AIC corresponds to the Gaussian copula which means that it has the best copula function for this study (Yazdani and Albrecht 1987).

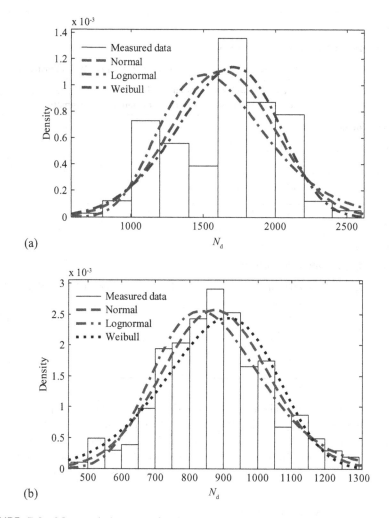

FIGURE 7.8 Measured data and fitted probability models of N_d: (a) BO-SG-01; (b) BO-SG-02.

7.3.4.2 Dimension Gaussian Copula Function

There are two variables which have correlation in the joint function. Equation (7.11) can be rewritten as follows:

$$C\left(u_1, u_2, \theta\right) = \int_{-\infty}^{\Phi^{-1}(u_1)} \int_{-\infty}^{\Phi^{-1}(u_2)} \frac{1}{2\pi\sqrt{1-\theta^2}} \exp\left[-\frac{s^2 - 2\theta st + t^2}{2\left(1-\theta^2\right)}\right] ds dt \qquad (7.15)$$

where u_1 and u_2 are the marginal distribution models of S_{eq} and N_d. θ is the correlation coefficient between S_{eq} and N_d. The θ were calculated by Equations (7.13) and (7.14)

TABLE 7.3

Parameters for Models of S_{eq} and N_d

Models		Normal distribution		Lognormal distribution		Weibull distribution	
		Mean value	Standard deviation	Mean value	Standard deviation	Scale parameter	Shape parameter
Parameters of models		μ	σ	μ	σ	a	b
B0-SG-01	S_{eq}(MPa)	10.10	3.49	2.25	0.38	11.31	3.31
	N_d	1634	359	7.37	0.24	1776	5.4
B0-SG-01	S_{eq}(MPa)	7.2	1.08	7.2	1.3	7.23	1.36
	N_d	814	155	6.75	0.18	939	6.14

TABLE 7.4

Parameters of Copula Functions

Types of copulas	θ/ρ (Detail 1)	V_{AIC} (Detail 1)	θ/ρ (Detail 2)	V_{AIC} (Detail 2)
Gaussian	0.58	−125	0.25	−89
t-copula	0.51	−101	0.20	−51
Gumbel	0.18	−57	0.11	−35
Frank	5.2	−77	2.4	−66
Clayton	5.4	−81	2.0	−75

and the θ values for B0-SG-01 and B0-SG-02 are 0.58 and 0.25, respectively. The joint function probability density distributions are shown in Figure 7.9. To analyze the adaptability of the joint function, 1000 simulated samples were compared with measured data as shown in Figure 7.10. The measured data and simulated data have a similar distribution which demonstrates the joint model's applicability.

7.3.5 FATIGUE RELIABILITY INDEX

If all the variables in Equation (7.5) are independent, the expression of fatigue reliability index can be written as follows:

$$\beta = \frac{\mu_{\ln \Delta} - \left[\mu_{\ln e} + \ln(N_d) - \mu_{\ln A} + \mu_{\ln S_{eq}}\right]}{\sqrt{\ln\left[\left(1+\delta_\Delta^2\right)\cdot\left(1+\delta_A^2\right)\cdot\left(1+\delta_e^2\right)\right] + B^2 \ln(1+\delta_{S_{eq}}^2)}} \tag{7.16}$$

where δ is a coefficient of variation, $\delta = \sigma/\mu$.

The values of fatigue reliability index are more accurate because the variable's correlation is considered. However, the computing process of fatigue reliability index is more complicated and it is difficult to obtain analytic solutions because of the

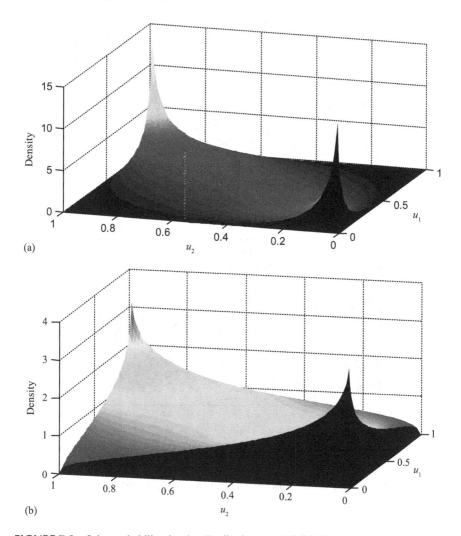

FIGURE 7.9   Joint probability density distributions: (a) B0-SG-01; (b) B0-SG-02.

implicit performance joint functions. The fatigue reliability index calculation process can be divided into three steps. The first step is to sample the correlation matrix (S_{eq} and N_d) and independence matrix ($\mathbf{A}$, Δ, and $\mathbf{e}$) based on the joint function model and independence probability models. The larger number of samples makes the computed result for the matrix more accurate. Frangopol et al. (2008) gives the following recommendation for N:

$$N > \frac{100}{P_f} \tag{7.17}$$

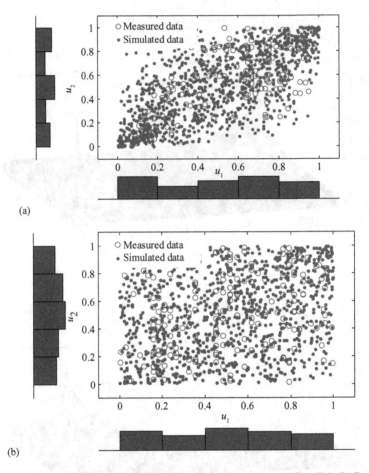

FIGURE 7.10   Scatter plots of measured data and simulated data: (a) Detail 1; (b) Detail 2.

The second step is to calculate Equation (7.7) based on the samples, which can then be rewritten as

$$\mathbf{P} = \varDelta - \mathbf{e} \cdot \frac{365 \cdot m \cdot \int_0^{T_m} (1+\alpha)^m \, dm \cdot \mathbf{N_d} \cdot \mathbf{S}_{eq}^B}{\mathbf{A}} \tag{7.18}$$

where the matrix $\mathbf{P}$ is the value of computation.

The third step is to select N_f. The number of variables N_f from $\mathbf{P}$ and $P < 0$ was accumulated. The expression of fatigue failure probability is as follows:

$$p_f = \frac{N_f}{N} \tag{7.19}$$

The reliability index can be expressed as

$$\beta = \Phi^{-1}\left(1 - P_{\mathrm{f}}\right) \tag{7.20}$$

It is important to note that the Monte Carlo method can only calculate the value of β less than 6. If $\beta > 6$, the number of samples required is more than 10^9, which is difficult for a standard computer to handle.

The fatigue reliability index, β, reflects the condition of fatigue damage. Using Equation (7.16) and the Monte Carlo method to analyze the reliability index considering the correlation of variables, β is calculated for different service years. Figure 7.11(a) compares the time-variant reliability index β accounting for variables correlation for Detail 1. It can be seen that when considering the variables correlation, β is lower than the value of independent variable. Both values of β are lowered to the target minimum $\beta_{\mathrm{Tagret}} = 2$ within 200 years. Figure 7.11(b) compares the

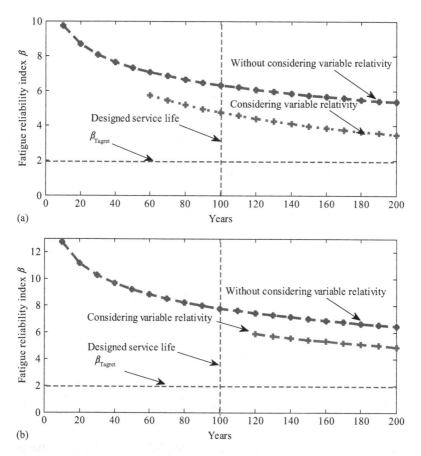

FIGURE 7.11 Comparisons of time-variant reliability index β with and without influence of temperature: (a) Detail 1; (b) Detail 2.

time-variant reliability index β with and without considering variables correlation for Detail 2; the value of reliability without considering variables correlation is higher than that when considering variable correlation. Both values always maintain a higher lever during the design life. After 200 service years, the reliability index of Detail 2 is still higher than β_{Tagret}.

The traffic growth rate α is another factor which affects fatigue reliability. Figure 7.12(a) shows the reliability for Detail 1 with $\alpha = 2\%$ and 5%. It is observed that the reliability decreases to the target minimum β_{Tagret} after 67.2 and 94.8 service years, respectively. Figure 7.12(b) shows the reliability for Detail 2 with $\alpha = 2\%$ and 5%. It is observed that the target minimum β_{Tagret} corresponds to the service times of 120 years and 82.3 years, respectively. For Detail 2 when $\alpha = 0\%$, β is still higher than β_{Tagret} after 200 service years.

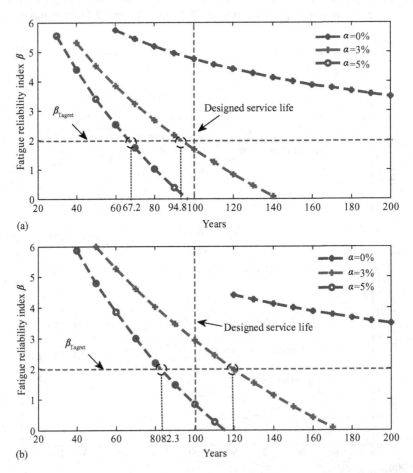

FIGURE 7.12 Comparisons of time-variant reliability index β with different traffic growth rates accounting for: (a) traffic growth rate for Detail 1; (b) traffic growth rate for Detail 2.

7.4   CONCLUSIONS

This chapter presents an approach for steel box weld fatigue reliability considering variables correlation using copula based on SHM data. The conclusions drawn from this study are as follows:

1. The flowchart of fatigue reliability assessment considering variables correlation was proposed in this paper using copula based on strain data. Five types of variables correlation were analyzed and discussed. The statistical data indicate that there is a distinct correlation between daily equivalent fatigue stress, S_{eq}, and stress cycle, N_d.
2. Five types of copula functions were introduced to establish the joint function for reliability limit equation. The Gaussian copula was selected as the best function for the Nanxi Yangtze River Bridge statistical data by the AIC method.
3. The fatigue reliability with or without considering variables correlation for details of rib-to-deck and bottom of rib was assessed. The value of reliability without considering variables correlation is higher than that considering variable correlation. Both values always maintain a higher lever during the design life. β is lowered to the target minimum β_{Tagret} after 67.2 and 94.8 service years with a traffic growth rate value of 2% and 5%.

REFERENCES

AASHTO. 2002. *Standard specifications for highway bridges*. Washington, DC: American Association of State Highway and Transportation Officials.

Akaike, H. 1974. A new look at the statistical model identification. *IEEE Transactions on Automatic Control* 19(6): 716–723.

Ang, A. H.-S. and Tang, W. 2007. *Probability concepts in engineering: emphasis on application to civil and environmental engineering*. 2nd ed. New York: John Wiley and Sons.

Dan, M. F. 2008. Probability concepts in engineering: emphasis on applications to civil and environmental engineering. *Structure and Infrastructure Engineering* 4(5): 413–414.

Dan, M. F., Strauss, A. and Kim, S. 2008. Bridge reliability assessment based on monitoring. *Journal of Bridge Engineering* 13(3): 258–270.

Deng, Y. and Ding Y. 2011. Fatigue reliability assessment for bridge welded details using long-term monitoring data. *Science China Technological Sciences* 54(12): 3371–3381.

Downing, S. D. and Socie, D. F. 1982. Simple rainflow counting algorithms. *International Journal of Fatigue* 4(1): 31–40.

Eryilmaz, S. 2014. Multivariate copula based dynamic reliability modeling with application to weighted-k -out-of- n, systems of dependent components. *Structural Safety* 51(51): 23–28.

Fengler, M. R. and Okhrin, O. 2014. Managing risk with a realized copula parameter. *Computational Statistics and Data Analysis* 100:131–152.

Frangopol, D. M., Strauss, A. and Kim, S. (2008). Bridge reliability assessment based on monitoring. *Journal of Bridge Engineering*, *13*(3), 258–270.

Goda, K. 2010. Statistical modeling of joint probability distribution using copula: Application to peak and permanent displacement seismic demands. *Structural Safety* 32(2): 112–123.

Guo, T. and Chen, Y. W. 2011. Field stress/displacement monitoring and fatigue reliability assessment of retrofitted steel bridge details. *Engineering Failure Analysis* 18(1): 354–363.

Guo, T., Frangopol, D. M. and Chen, Y. 2012. Fatigue reliability assessment of steel bridge details integrating weigh- in- motion data and probabilistic finite element analysis. *Computers and Structures* 112–113(4): 245–257.

Jiang, C. et al. 2014. Structural reliability analysis using a copula-function-based evidence theory model. *Computers and Structures* 143: 19–31.

Keissar, K. et al. 2010. Non-invasive baroreflex sensitivity assessment using wavelet transfer function-based time-frequency analysis. *Physiological Measurement* 31(7): 1021–1036.

Kwon, K. and Dan, M. F. 2010. Bridge fatigue reliability assessment using probability density functions of equivalent stress range based on field monitoring data. *International Journal of Fatigue* 32(8): 1221–1232.

Li, D. Q. et al. 2012. Uncertainty analysis of correlated non-normal geotechnical parameters using Gaussian copula. *Science China Technological Sciences* 55(11): 3081–3089.

Liu, M., Frangopol, D. M. and Kwon, K. 2010. Fatigue reliability assessment of retrofitted steel bridges integrating monitored data. *Structural Safety* 32(1): 77–89.

Liu, Y., Lu, N. W. and Yin, X. F. 2016. A hybrid method for structural system reliability-based design optimization and its application to trusses. *Quality and Reliability Engineering*, 32(2): 595–608.

Liu, Y., Zhang, H. Deng, Y. and Jiang, N. 2017a. Effect of live load on simply supported bridges under a random traffic flow based on weigh-in-motion data. *Advances in Structural Engineering*, 20(5), 722–736.

Liu, Y., Zhang, H., Li, D., Deng, Y. and Jiang, N. 2017b. Fatigue reliability assessment for orthotropic steel deck details using copulas: application to Nan-Xi Yangtze River Bridge. *Journal of Bridge Engineering*, 23(1), 04017123.

Miner, M. 1945. Cumulative damage in fatigue. *Journal of Applied Mechanics* 12(3): 159–164.

Nelsen, R. B. 2006. *An introduction to Copulas*. New York: Springer Series in Statistics.

Ni, Y. Q., Ye, X. W. and Ko, J. M. 2010. Monitoring-based fatigue reliability assessment of steel bridges: analytical model and application. *Journal of Structural Engineering* 136(12): 1563–1573.

Noh, Y., Choi, K. K. and Liu, D. 2009. Reliability-based design optimization of problems with correlated input variables using a Gaussian Copula. *Structural and Multidisciplinary Optimization* 38(1): 1–16.

Nur, Y. and Albrecht, P. 1987. Risk analysis of fatigue failure of highway steel bridges. *Journal of Structural Engineering* 113(3): 483–500.

Singh, V. P. and Zhang, L. 2006. Bivariate flood frequency analysis using the Copula method. *Journal of Hydrologic Engineering* 11(2): 150–164.

Sklar, M. Fonctions de Répartition À N (1960). Dimensions et leurs marges. *Publications de l'Institut de Statistique de l'Université de Paris* 8: 229–231.

Stefano, D. and McNeil, A. J. 2005. The t Copula and related Copulas. *International Statistical Review* 73(1): 111–130.

Tang, X. S. et al. 2013a. Bivariate distribution models using copulas for reliability analysis. *Proceedings of the Institution of Mechanical Engineers Part O Journal of Risk and Reliability* 227(5): 499–512.

Tang, X. S. et al. 2013b. Impact of copulas for modeling bivariate distributions on system reliability. *Structural Safety* 44(2334): 80–90.

Tang, X. S. et al. 2015. Copula-based approaches for evaluating slope reliability under incomplete probability information. *Structural Safety* 52: 90–99.

Tong, G., Dan, M. F. and Chen, Y. 2012. Fatigue reliability assessment of steel bridge details integrating weigh-in-motion data and probabilistic finite element analysis. *Computers and Structures* 112(4): 245–257.

Tong, G., Li, A. and Li, J. 2008. Fatigue life prediction of welded joints in orthotropic steel decks considering temperature effect and increasing traffic flow. *Structural Health Monitoring* 7(3): 189–202.

Wirsching, P. H. 1984. Fatigue reliability for offshore structures. *Journal of Structural Engineering* 110(10): 2340–2356.

Wu, X. Z. 2014. Assessing the correlated performance functions of an engineering system via probabilistic analysis. *Structural Safety*. 52:10–19.

Yazdani, N. and Albrecht, P. (1987). Risk analysis of fatigue failure of highway steel bridges. *Journal of Structural Engineering*, 113(3): 483–500.

Zhao, Z. W. et al. 1994. Fatigue-reliability evaluation of steel bridges. *Journal of Structural Engineering* 120(5): 1608–1623.

8 Fatigue Crack Propagation of Rib-to-deck Double-sided Welded Joints of Orthotropic Steel Bridge Decks

Yang Liu

Hunan University of Technology, China

Fanghui Chen

Changsha University of Science and Technology, China

Naiwei Lu

Changsha University of Science and Technology, China

CONTENTS

8.1 INTRODUCTION

Orthotropic steel decks (OSDs) are commonly used in cable-stayed, suspension, and other bridge structures which require high strength combined with light weight (Lu et al. 2017a; Mao et al. 2019). However, the fatigue durability of OSDs encounters a challenge due to the aging of the structural materials combined with increasing traffic loads (Lu et al. 2019; Zhang et al. 2020). Several authors have mentioned that fatigue is one of the main causes of structural failure (Deng et al. 2016, Song et al. 2016; Lu et al. 2017b) and that it should be the primary consideration to extend the service life of a bridge. Based on a review on fatigue cracks occurring in OSDs, different causes of fatigue cracking were distinguished (Chitoshi 2006; Kolstein 2007), and the authors generally agree with the following categories: (1) Cracking induced by welding crack-like defects: fatigue cracks initiate from welding defects introduced during manufacturing; (2) Load-induced cracking: fatigue cracking due to low fatigue strength of the connection details; and (3) Distortion-induced cracking: fatigue cracks initiation or growth due to out-of-plane stresses and deformations under wheel loads.

In recent years, the rib-to-deck welded joints of OSDs have received significant attention due to fatigue cracking problems. Fatigue cracks initiating at the weld root of rib-to-deck welded joints are one of the most common types of cracks observed (Kainuma et al. 2016). Recently, even long through-roof type fatigue cracks have been detected, and the frequency of occurrence of such cracks is gradually increasing (Kainuma et al. 2016; Ya et al. 2011). These fatigue cracks are the most dangerous as their identification is difficult during inspection due to the hidden locations, unless the resulting damage of the wear layer becomes visible (Wang et al. 2019b). Several authors have mentioned that weld penetration is one of the main influencing factors affecting the fatigue performance of rib-to-deck single-sided welded joints. Some studies have shown that increasing the weld penetration has a positive impact on improving the fatigue performance of rib-to-deck single-sided welded joints (Dung et al. 2014; Fu et al. 2017), while other studies have shown that a shallower weld penetration has a positive effect on improving the fatigue performance of rib-to-deck single-sided welded joints (Kainuma et al. 2016; Ya et al. 2011). The effect of weld penetration on fatigue performance of rib-to-deck single-sided welded joints of OSDs is still inconclusive. Therefore, more careful consideration is needed for the design of rib-to-deck welded joints to overcome the disadvantages of single-sided welding.

More recently, various innovative rib-to-deck details have been proposed, which are expected to improve the anti-fatigue performance of rib-to-deck welded joints. Masahiro et al. (2014) proposed the use of double-sided welding for the first time to attach the closed ribs to the deck plates in OSDs, and conducted tests to investigate the effect of the inside fillet weld on fatigue durability of rib-to-deck welded joints. You et al. (2018) conducted a series of comparative tests on the fatigue performance of double- and single-sided welding of rib-to-deck welded joints. The results indicated that double-sided welds could significantly improve the fatigue life of rib-to-deck welded joints. Liu et al. (2019) studied the parameter analysis of the influence of weld penetration on the fatigue performance of rib-to-deck double-sided welded joints and the results indicated that the penetration exhibited very little effect. However, currently the research on fatigue characteristics of rib-to-deck double-sided welded joints is still lacking.

The linear elastic fracture mechanics (LEFM) has been used to understand the fatigue behavior of rib-to-deck welded joints of OSDs. Wu et al. (2019) analyzed the fatigue crack propagation of rib-to-deck single-sided welded joints in OSDs by an LEFM of multi-point extrapolation. Wang et al. (2019a) employed a finite element method (FEM) to simulate the crack propagation of rib-to-deck single-sided welded joints, and validated it against fatigue experiments. Wang et al. (2016) employed the extended FEM to simulate the fatigue crack growth of rib-to-deck welded joints. Previous studies have mainly dealt with two-dimensional (2D) problems to simulate crack growth of rib-to-deck welded joints. In fact, the defects of rib-to-deck welded joints are subjected to mixed mode loading under the vehicle load. Therefore, better understanding of the mixed mode fatigue crack propagation behavior of the rib-to-deck double-sided welded joints of OSDs is highly desirable.

This study investigated the mixed mode fatigue crack propagation behavior of the rib-to-deck double-sided welded joints of OSDs by LEFM. The 3D FE models of rib-to-deck double-sided welded joints were established to determine the SIF related to the three modes of fracture. Mixed mode crack growth behaviors were analyzed in terms of the SIF variation along the crack front, crack growth paths, and crack shape variation from which the fatigue life of the weld toe of rib-to-deck welded joints was predicted. Finally, the influences of initial crack size and aspect ratio (a/c) a/c) on fatigue life were analyzed.

8.2 DETAILS OF DOUBLE-SIDE WELDED JOINTS

The more advanced welding technology, namely, U-rib internal welding technology, was adopted to add fillet welds inside the closed longitudinal ribs to form rib-to-deck double-sided welded joints. Compared to the traditional single-sided welded joints, the innovative double-sided welded joints are expected to improve the fatigue performance of rib-to-deck. Recently, the rib-to-deck double-sided welded joints have been successfully applied in the construction of actual bridges in China, such as Zhuankou Yangtze River Bridge and Jiayu Yangtze River Bridge. The details of rib-to-deck double-sided welded joints are shown in Figure 8.1.

8.3 FATIGUE CRACK GROWTH SIMULATION AND LIFE PREDICTION METHOD

8.3.1 M-Integral for SIF Determination

The M-Integral, which is a numerical method for the calculation of SIF related to the three modes of fracture, was first proposed by Yau et al. (1980) from the J-Integral (Rice, 1968). The J-Integral is defined in terms of

$$J = \int_\Gamma \left(\sigma_{ij} \frac{\partial u_i}{\partial x_1} - W \delta_{1j} \right) \frac{\partial q}{\partial x_j} ds \tag{8.1}$$

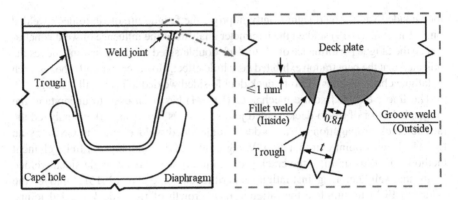

FIGURE 8.1   Details of rib-to-deck double-sided welded joints.

where Γ is the contour integration around the crack tip, u_i is the displacement vector, $dsds$ is the increment of arc length along Γ, x_1 is the position vector of the integration point, $\delta_{1j}\delta_{1j}$ is the crack tip opening displacement, and q is a function that is 1 at the crack tip and 0 on the boundary of the integration domain; the strain energy density W is defined as $W = (\sigma_{ij}\,\varepsilon_{ij})/2$ $W = \dfrac{1}{2}\sigma_{ij}\varepsilon_{ij}$, σ_{ij} is the stress tensor, and ε_{ij} is the strain tensor.

For linear analysis, the field variables associated with the two solutions are assumed and superimposed (Yau et al. 1980), and defined as follows:

$$\begin{cases} \sigma_{ij} = \sigma_{ij}^{(1)} + \sigma_{ij}^{(2)} \\ \varepsilon_{ij} = \varepsilon_{ij}^{(1)} + \varepsilon_{ij}^{(2)} \\ u_{ij} = u_{ij}^{(1)} + u_{ij}^{(2)} \end{cases} \tag{8.2}$$

The *J*-integral superposed state can be defined (Yau et al. 1980) as:

$$J = J^{(1)} + J^{(2)} + M^{(1,2)} \tag{8.3}$$

where

$$= \int_{\Gamma}\left(\sigma_{ij}^{(1)}\frac{\partial u_i^{(1)}}{\partial x_1} - W^{(1)}\delta_{1j} \right)\frac{\partial q}{\partial x_j}\,ds \tag{8.4}$$

$$J^{(2)} = \int_{\Gamma}\left(\sigma_{ij}^{(2)}\frac{\partial u_i^{(2)}}{\partial x_1} - W^{(2)}\delta_{1j} \right)\frac{\partial q}{\partial x_j}\,ds \tag{8.5}$$

$$M^{(1,2)} = \int_{\Gamma}\left(\sigma_{ij}^{(1)}\frac{\partial u_i^{(2)}}{\partial x_1} + \sigma_{ij}^{(2)}\frac{\partial u_i^{(1)}}{\partial x_1} - W^{(1,2)}\delta_{1j} \right)\frac{\partial q}{\partial x_j}\,ds/A_q \tag{8.6}$$

where $A_q = \int_L q_t \, ds$, q_t is the value of the q function along the crack front and $W^{(1,2)}$ is the interaction strain energy density, defined by

$$W^{(1,2)} = \sigma_{ij}^{(1)}\varepsilon_{ij}^{(2)} = \sigma_{ij}^{(2)}\varepsilon_{ij}^{(1)} \tag{8.7}$$

and $M^{(1,2)}$ is the M-Integral.

For small scale yielding, the energy release rate G is equal to the J- Integral (Lv et al. 2018).

$$G = J = \frac{1-v^2}{E}K_I^2 + \frac{1-v^2}{E}K_{II}^2 + \frac{1+v}{E}K_{III}^2 \tag{8.8}$$

where v is Poisson's ratio, E is Young's modulus, and K_i is SIF, $K_i = K_i^{(1)} + K_i^{(2)}, i = I, II, III$.

The relationship between M-Integral in terms of material properties and the SIF is (Lv et al. 2018)

$$M^{(1,2)} = 2 \times \left[\frac{1-v^2}{E}K_I^{(1)}K_I^{(2)} + \frac{1-v^2}{E}K_{II}^{(1)}K_{II}^{(2)} + \frac{1+v}{E}K_{III}^{(1)}K_{III}^{(2)} \right] \tag{8.9}$$

Moreover, the two definitions for the M-Integral are equal.

$$\int_\Gamma \left(\sigma_{ij}^{(1)} \frac{\partial u_i^{(2)}}{\partial x_1} + \sigma_{ij}^{(2)} \frac{\partial u_i^{(1)}}{\partial x_1} - W^{(1,2)}\delta_{1j} \right) \frac{\partial q}{\partial x_j} \, ds/A_q$$

$$= 2 \times \left[\frac{1-v^2}{E}K_I^{(1)}K_I^{(2)} + \frac{1-v^2}{E}K_{II}^{(1)}K_{II}^{(2)} + \frac{1+v}{E}K_{III}^{(1)}K_{III}^{(2)} \right] \tag{8.10}$$

K_I, K_{II}, K_{III} can be calculated by using Equation (8.10) in FE analysis.

8.3.2 MIXED MODE SIF RANGE

In most engineering cases a mixed mode of modes I, II, and III of fracture can be found. The equivalent SIF K_{eq} K_{eq} of the mixed mode can be calculated by using Equation (8.11) from BS 7910 (2005). The equivalent SIF range ΔK_{eq} is calculated by

$$K_{eq} = \sqrt{K_I^2 + K_{II}^2 + \frac{K_{III}^2}{1-v}} \tag{8.11}$$

$$\Delta K_{eq} = K_{eq,max} - \max\left(K_{eq,min}, 0\right) \tag{8.12}$$

where v is Poisson's ratio; K_I, K_{II}, K_{III} are separate SIF for all three modes of fracture, respectively; and $K_{eq,max}$ $K_{eq,\,max}$ and $K_{eq,min}$ $K_{eq,\,min}$ are the maximum and minimum values of K_{eq}, respectively.

8.3.3 KINK ANGLE MODEL

The direction of crack propagation is determined by the kink angle. The maximum tensile stress (MTS) theory proposed by Erdogan and Sih (1963) is one of the most popular theories. It assumes that the crack extension direction θ_c is consistent with the maximum value of the hoop stress. For materials with isotropic properties, the crack extension direction θ_c is given by Liu et al. (2020)

$$\theta_c = 2\arctan\left[\frac{-2\left(K_{II}/K_I\right)}{1+\sqrt{1+8\left(K_{II}/K_I\right)^2}}\right] \tag{8.13}$$

8.3.4 CRACK EXTENSION TYPE

The specified median crack front extension is used to grow and extend the crack, which specifies the median crack growth increments over each growth step. The crack extension at the crack front point i, Δa_i is computed by using Equation (8.14) (Liu et al. 2016). The specified crack extension is illustrated in Figure 8.2.

$$\Delta a_i = \Delta a_m \left(\frac{\Delta K_{eq,i}}{\Delta K_{eq,m}}\right)^n \tag{8.14}$$

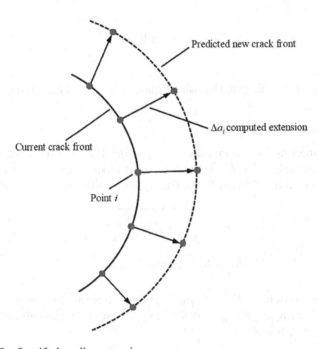

FIGURE 8.2 Specified median extension.

where Δa_m is the crack extension at the crack front point, $\Delta K_{eq,\,i}$ is the equivalent SIF range at the crack front point, $\Delta K_{eq,\,m}$ is the median equivalent SIF range, and n is the material parameter.

8.3.5 FATIGUE LIFE PREDICTION

The Paris law (1963) was modified to predict the crack growth rate of the mixed mode, which can be expressed as:

$$\frac{da}{dN} = C\left(\Delta K_{eq}\right)^n \tag{8.15}$$

$$N = \int_{a_0}^{a_f} \frac{da}{C\left(\Delta K_{eq}\right)^n} \tag{8.16}$$

where a is crack length and N is the number of cycles due to which a crack propagates from its initial crack length, a_0, to a final crack length, and a_f can be expressed as Equation (8.16). C and n are empirical coefficients, which are specified along with the values for the SIF threshold range ΔK_{th}.

8.3.6 STEP-WISE PROCEDURE

In this study, FRANC3D (2016) and ABAQUS (2019) were used to simulate mixed mode fatigue crack growth. The typical flowchart is shown in Figure 8.3.

8.3.7 NUMERICAL EXAMPLES AND VERIFICATION

A 3D mixed mode fatigue crack growth model was used to simulate the surface crack in the T-joint under constant amplitude loading, giving a comparative analysis of the calculated number of load cycles to the experimental results provided by Nikfam et al. (2019). ASTM-572 steel welded T-joint specimens and initial crack geometry are shown in Figure 8.4. The values of Young's modulus, ultimate tensile stress, yield stress, and Poisson's ratio are 207 GPa, 572 MPa, 406 MPa, and 0.3, respectively. Load amplitude of $P = 30$ kN was used for loading. FRANC3D and ABAQUS were used to simulate the mixed mode fatigue crack growth of a T-joint. The T-joint was modeled by using 8-nodes solid elements (C3D8R). The mesh size of the global model is 10 mm, and that of the cracked sub-model is 0.02 mm (Figure 8.4). The Paris law constants for ASTM-572 steel were taken as $nn = 2.88$ and $C = 6.77 \times 10^{-13}$ (MPa and mm) from BS 7910 (2005).

Figure 8.5 presents the comparison between the simulation results and the two experimental results by Nikfam et al. (2019). The comparative analysis indicates that the numerical simulation results are in sound agreement with the experimental data. The number of cycles of numerical simulation is very close to the mean number of cycles of two experiments, with a difference of 2.1%. Noteworthy it that the error in the crack growth simulation may be caused by the initial crack geometry. In the experiments, the initial crack geometry was not a perfect semi-elliptical shape.

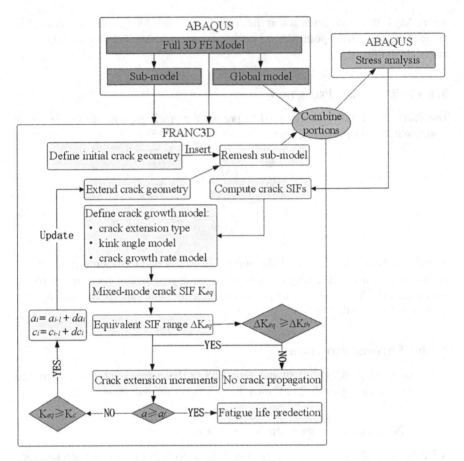

FIGURE 8.3 Flowchart for simulation of crack growth.

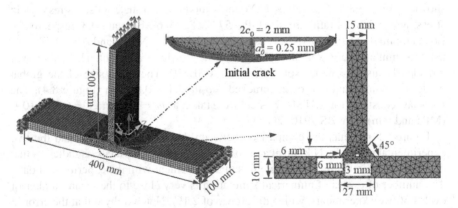

FIGURE 8.4 Surface crack in plate under tension.

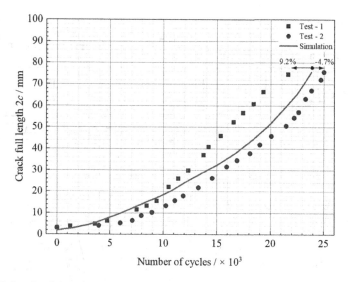

FIGURE 8.5   Crack length versus number of cycles.

Therefore, it is feasible to simulate mixed mode crack growth of 3D problem by using FRANC3D and the results are reasonably accurate.

8.4   CASE STUDY

8.4.1   BACKGROUND OF PROTOTYPE BRIDGE

Jiayu Yangtze River Bridge which is located in Hubei, China and has a main span of 920 m was considered as the research subject. It is an unsymmetrical composite girder cable-stayed bridge. The cross-section of the main girder is illustrated in Figure 8.6. The width of the main girder is 33.5 m and the standard height is 3.0 m. The deck plates are designed to be 16–24 mm thick and the longitudinal closed U-ribs are 8 mm thick. The ribs are 300 mm high and 300 mm wide on top and have a width of 180 mm at the lower soffit. The distance between the longitudinal ribs is equal to 300 mm. The U-rib and deck plate were connected by double-sided welding. First, a small intelligent welding robot was used to weld the fillet weld inside the joint, and then the conventional welding technology was employed to weld the groove fillet weld with 80% penetration outside the joint. The clearance between the U-rib and the deck plate assembly was no more than 0.5 mm.

8.4.2   FINITE ELEMENT MODEL

To study the mixed mode fatigue crack propagation behavior of rib-to-deck double-side welded joints of OSDs, two different 3D FE models were established. The first FE model was used in a stress analysis to identify the most adverse situation for loading, which was named as no-crack FE model. The second FE model was used in a

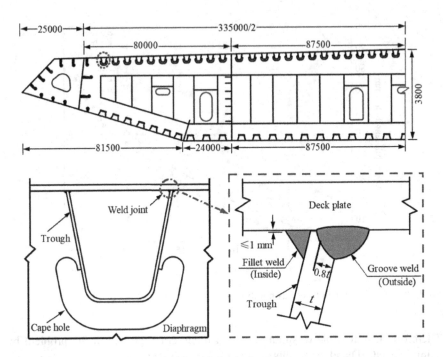

FIGURE 8.6   Cross-section of steel box girder (unit: mm).

fracture analysis and was named as cracked FE model. The no-crack FE model, which consists of five longitudinal U-ribs and three diaphragms as shown in Figure 8.7, was established by using ABAQUS with C3D8R. Young's modulus of 206 GPa and Poisson's ratio of 0.3 were used in the FE model to represent material properties for steel Q345qD. The following boundary conditions were preassigned in ABAQUS: (1) The vertical translation (Y direction) and rotations (X and Z directions) are constrained for all nodes of the lower edge of the diaphragm to simulate support of the diaphragm; (2) The longitudinal translation (Z direction) and rotations (X and Y directions) are constrained for all nodes of the two ends of the model to simulate the boundary conditions at the interior rib and deck plate; and (3) The transverse translation (X direction) and rotations (Y and Z directions) are constrained for all nodes of the two sides of the model to simulate the boundary condition at the interior diaphragm and deck plate. The cracked model was established by using FRANC3D, by inserting fatigue cracks into the no-crack model. In order to optimize the computational cost associated with the process, the uncracked sub-model, shown in Figure 8.7 as local, was selected and imported into FRANC3D. A semi-elliptic initial crack was inserted into the uncracked sub-model and then remeshing was carried out by using FRANC3D to establish a cracked sub-model, as shown in Figure 8.7. The cracked sub-model was connected to the global model by merging nodes. In this way, it could be subdivided to avoid remeshing of the global model at each step of crack growth and to favor the transition of the finer mesh (i.e., cracked sub-model mesh) to the coarser mesh (i.e., global mesh). The mesh size of the global model was 10 mm, and that of the cracked sub-model was 0.02 mm.

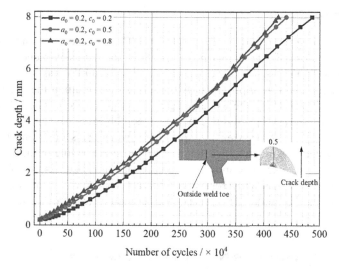

FIGURE 8.7   FE model of rib-to-deck welded joints (unit: mm).

8.4.3   GEOMETRY AND LOCATION OF INITIAL FLAWS

The initial crack geometry is an important parameter that directly affects fatigue crack propagation. In general, fatigue cracks grow irregularly in a non-uniform stress field (Gadallah et al. 2017). For the sake of simplicity, a semi-elliptical surface initial crack with a depth of 0.2 mm and full surface length of 1 mm was inserted into the no-crack sub-model. Moreover, four types of cracks, namely, Crack I, Crack II, Crack III, and Crack IV as shown in Figure 8.8, were assumed for fatigue analysis.

Crack I: the outside toe-deck, which initiates at the outside weld toe and propagates into the deck.

Crack II: the outside root-deck, which initiates at the outside weld root and propagates into the deck.

Crack III: the inside root-deck, which initiates at the inside weld root and propagates into the deck.

Crack IV: the inside toe-deck, which initiates at the inside weld toe and propagates into the deck.

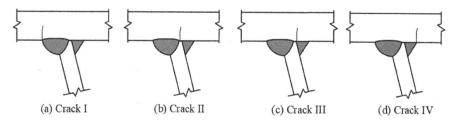

(a) Crack I   (b) Crack II   (c) Crack III   (d) Crack IV

FIGURE 8.8   Schematic illustration of different forms of crack.

8.4.4 FATIGUE LOAD

The fatigue load model III in EN 1991–2 (2003) was used for simulation. The geometry of mode III is shown in Figure 8.9. The stress influence line in the transverse direction of rib-to-deck welded joints of OSD was short under the wheel loading, and was approximately three times the width of the U-rib opening (Xiao et al. 2008), while the wheel spacing of single axles in mode III was 2 m. Therefore, the stress interference between neighboring wheels of single axle could be neglected. Moreover, the spacing between the second and third axles was 6 m, which was much larger than the stress influence line in the longitudinal direction. For the sake of simplicity, a fatigue load mode consisting of two axles, each having one wheel and each with a contact surface of 400 mm by 400 mm was used in the FE model. The geometry of computational fatigue load is shown in Figure 8.9.

8.4.5 STRESS ANALYSIS FOR NO-CRACK FINITE ELEMENT MODEL

The double-axles fatigue load was used to simulate the moving vehicle load as shown in Figure 8.9. The DLOAD of the ABAQUS user subroutine was employed to simulate the fatigue load movement on the bridge. The increment of the moving step in the transverse direction was 50 mm, and that in the longitudinal direction was 100 mm, forming an array of load cases of 73×37 to simulate the moving of fatigue load. Loading cases are shown in Figure 8.10, where TC is defined as transverse movement and LC is defined as longitudinal movement.

Fatigue cracks are more sensitive to the transverse tensile stress, which is the driving force of crack propagation (Shi et al. 2013). The longitudinal influence line of transverse stress at the transverse locations ($X = -100$ mm) of the wheel center is presented in Figure 8.11 (a). It is observed that the transverse stress reaches the maximum value (tensile stress) when the double-axle center of fatigue load is in the

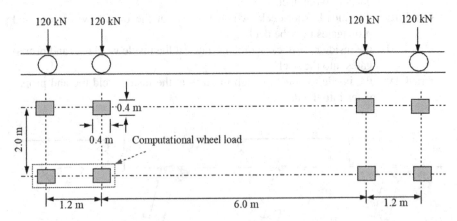

FIGURE 8.9 Fatigue load mode III and computational wheel load.

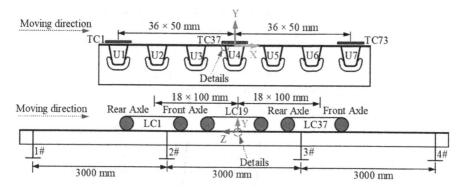

FIGURE 8.10 Loading cases: (a) transversal loading conditions; and (b) longitudinal loading conditions.

mid-span location ($Z = 0$ mm), which is the most adverse situation in the longitudinal direction. Furthermore, when the double-axle center of fatigue load is located in the mid-span ($Z = 0$ mm), the transverse influence line of transverse stress is obtained and presented in Figure 8.11 (b). Clearly, when the transverse position of vehicle load is ($X = -100$ mm), the maximum transverse stress (tensile stress) of the details can be obtained. Therefore, the most adverse situation in the transverse direction was observed when the wheel center is located at the transverse position of ($X = -100$ mm).

8.4.6 STATIC CRACK SIF ANALYSIS

The heavy vehicle load is one of the main causes of fatigue of OSD (Lu et al. 2019). In this study, considering the influence of overloading of the vehicle, the weight of each axle of 240 kN was used to perform fatigue analysis at the most adverse load position of vehicle load. Furthermore, the SIFs related to the three modes of fracture for the four types of cracks of the rib-to-deck double-sided welded joints in OSDs were determined. Under the most unfavorable loading conditions, the SIF values (K_I, K_{II}, K_{III}) along the crack front were identified by the non-dimensional curvilinear coordinate system, as shown in Figure 8.12. It is observed that the fatigue cracks at the rib-to-deck double-sided welded joints were the I–II–III mixed mode, which is dominated by mode I. For the initial crack of the weld toe, the values of K_I at the midpoint of the crack front are lower than the value of K_I at the two ends of the crack front. However, for the initial crack of the weld root, the values of K_I of the midpoint of the crack at the weld root are higher than the value of K_I at the two ends of the crack front. This is attributed to the presence of the fatigue cracks in the non-uniform stress field.

Figure 8.13 shows the values of $\Delta K_{eq}\Delta K_{eq}$ of the crack front of the four types of cracks. The ΔK_{eq} of the crack front at the outside weld toe is not much different from that at the inside weld toe. However, the ΔK_{eq} of the crack front at the weld toe is

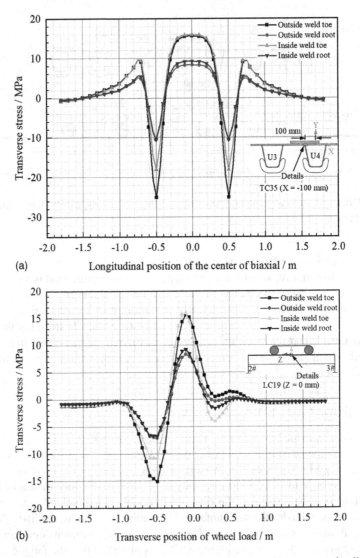

FIGURE 8.11   Stress-time curves: (a) longitudinal loading; and (b) transverse loading.

much greater than that at the weld root, which indicates that the fatigue failure of rib-to-deck double-sided welded joints is mainly determined by the fatigue crack of the weld toe. Notably, the maximum value of ΔK_{eq} of the crack front at welded root is 34.12 MPa·mm$^{1/2}$, which is far less than the threshold value of the SIF ($\Delta K_{th} = 63$ MPa·mm$^{1/2}$) in the International Institute of Welding (IIW) (Hobbacher 2008). Therefore, considering only the effect of vehicle load, the weld root of the rib-to-deck double-sided welded joints does not lead to crack growth. In this study, only the crack growth behavior of the weld toe of rib-to-deck double-sided welded joints was simulated.

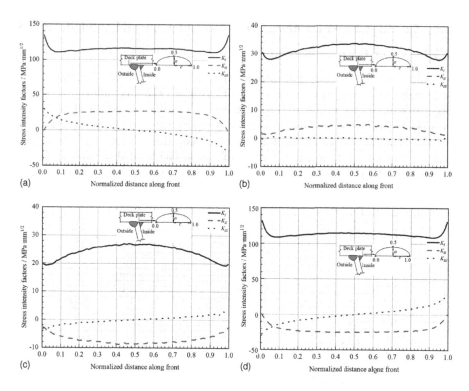

FIGURE 8.12 Static SIF of the initial crack: (a) Crack I, (b) Crack II, (c) Crack III, and (d) Crack IV.

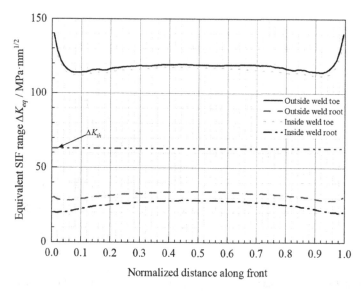

FIGURE 8.13 Equivalent SIF range of the initial crack.

8.5　CRACK GROWTH ANALYSIS AND FATIGUE LIFE PREDICTIONS

8.5.1　Crack Growth Behavior

The identification of crack propagation paths is essential for the comprehensive understanding of the fatigue crack growth behavior. The simulation results in a detailed visualization of the crack growth behavior of the welded toe of the rib-to-deck double-sided welded joints are shown in Figure 8.14, where various curves on the fatigue crack surface represent the crack propagation at each step. The results indicate that the crack growth behaviors of the outside welded toe and the inside welded toe of rib-to-deck double-sided joints are basically similar. During crack

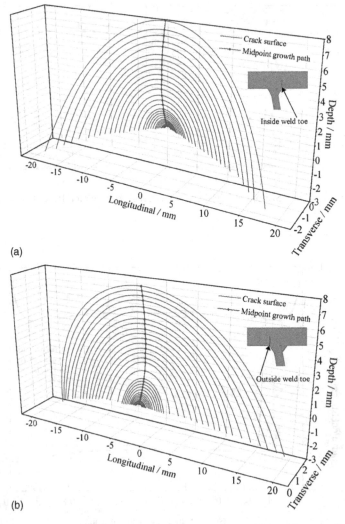

FIGURE 8.14　Spatial shape of crack propagation: (a) outside weld toe; and (b) inside weld toe.

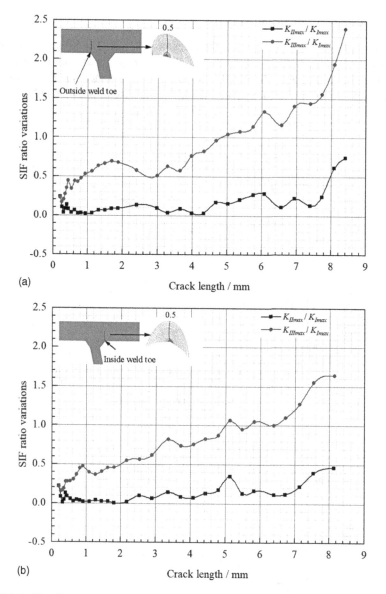

FIGURE 8.15 K_{IImax}/K_{Imax} and K_{IIImax}/K_{Imax} variations: (a) outside weld toe; and (b) inside weld toe.

propagation, the surface fatigue crack at the weld toe of rib-to-deck double-sided joints does not remain flat, but slightly deflected. This can be represented by the K_{IImax}/K_{Imax} and $K_{IIImax}/K_{Imax}K_{IImax}/K_{Imax}$ values, which quantify the contribution of mode II and mode III to the crack propagation, as shown in Figure 8.15. Moreover, it is also observed that the overall trend of the contribution of modes II and III increases after a certain number of crack growth steps. Therefore, the contribution of modes II and III cannot be neglected during the crack propagation.

8.5.2 VARIATION OF EQUIVALENT SIF RANGE

The variations of ΔK_{eq} directly affect the crack growth rate. The SIFs along the crack fronts of the welded toe of rib-to-deck double-sided joints at each stage of crack growth were determined. The variation of ΔK_{eq} with the crack length shown in the graphs refers to five propagation paths along the crack front as shown in Figure 8.16,

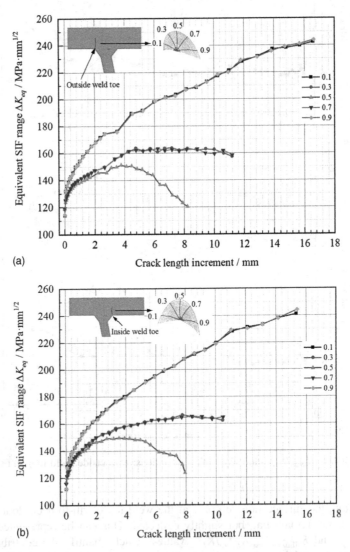

FIGURE 8.16 Variation curves of equivalent SIF range: (a) outside weld toe; and (b) inside weld toe.

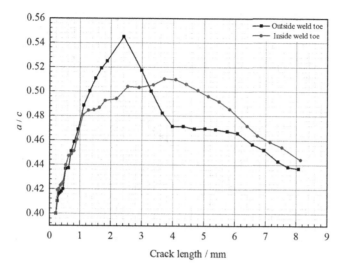

FIGURE 8.17 Variation curves of crack aspect ratio.

namely 0.1, 0.3, 0.5, 0.7, and 0.9, identified by using normalized coordinates. It is observed that the variation trends of ΔK_{eq} of the crack front of the outside weld toe and the inside weld toe are basically consistent during crack propagation. The ΔK_{eq} at the midpoint of the crack front along the propagation path (0.5) is always smaller than that of the crack front along the other propagation paths.

Noteworthy, the ΔK_{eq} is along the propagation path (0.5), which gradually increases and then decreases. Moreover, the results shown in Figure 8.17 illustrate the variation of fatigue crack aspect ratio (a/c) during crack propagation. It is observed that the a/c first increases and then decreases. This behavior is mainly due to the different crack growth rate along the front, that is, the crack growth rate at two ends of the crack front is higher than that at the midpoint of the crack front after a certain number of crack growth steps.

8.5.3 FATIGUE LIFE PREDICTIONS

The ΔK_{eq} value and the corresponding crack length estimated at a specific path along the crack front were determined. By using Equation (8.16), the number of load cycles of the rib-to-deck double-sided welded joints was calculated at each step, which is an average of the values computed for all mid-side nodes along the crack front when growing to the crack from one to the next. According to the recommended value of BS 7910 (2005), the Paris law's constants for Q345qD steel were taken as $n = 3$ and $C = 5.21 \times 10^{-13}$ (MPa and mm). According to the IIW (2008), it is assumed that structure fatigue failure occurs when the crack front is close to 50% of thickness of the deck plate. The curves of the fatigue life and the corresponding crack depth in the thickness direction of the deck plate are shown in Figure 8.18. The results reveal that

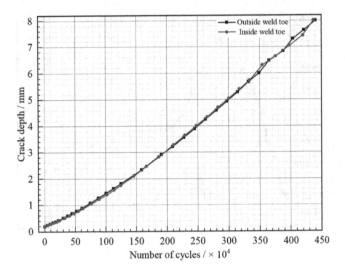

FIGURE 8.18   Fatigue crack depth versus number of cycles.

the value of the crack depth (aa = 8 mm) of the growth phase related to the crack with outside weld toe is equal to N = 4,400,093 cycles, and the value in the case with inside weld toe is equal to N = 4,367,796 cycles. Under the same fatigue load, the fatigue life of the inside welded toe is lower than that of the outside welded toe of rib-to-deck double welded joints.

8.6   DISCUSSION

As discussed previously, a surface semi-elliptical crack was assumed to be an initial defect to simulate crack growth. To investigate the impact of initial crack with a fixed aspect ratio (a_0/c_0 = 0.4) on fatigue life, the initial crack geometry was assumed to be a variable with three cases, including case 1 in which a_0 = 0.2 mm, c_0 = 0.5 mm; case 2 in which a_0 = 0.4 mm, c_0 = 0.1 mm; and case 3 in which a_0 = 1.0 mm, c_0 = 2.5 mm. The curves of crack depth versus the number of cycles were determined for three cases of outside welded toe as shown in Figure 8.19. The curves demonstrate that the initial crack geometry significantly affects the fatigue life.

Furthermore, in order to investigate the impact of initial crack aspect ratio (a_0/c_0) on fatigue life, the initial crack depth was assumed to be a constant a_0 = 0.2 mm, and the surface half-length c_0 was a variable changing between 0.2, 0.5, and 0.8 mm. The curves of crack depth versus the number of cycles were determined for three cases of outside welded toe as shown in Figure 8.20. Clearly, a_0/c_0 has a significant effect on fatigue life, and the fatigue life decreases significantly with a decrease in aspect ratio.

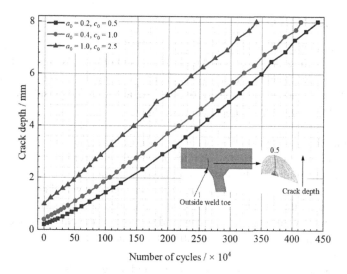

FIGURE 8.19 Crack depth versus number of cycles.

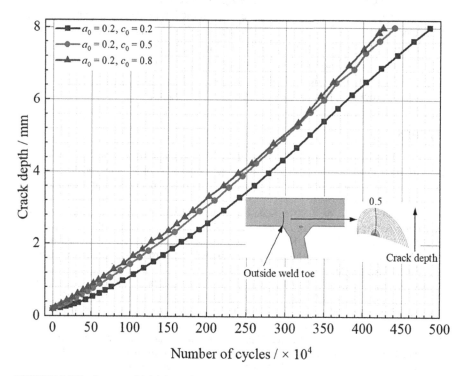

FIGURE 8.20 Impact of initial crack aspect ratio on crack depth.

8.7 CONCLUSIONS

This study presented a numerical simulation of 3D mixed mode fatigue crack growth behavior of rib-to-deck double-sided welded joints of OSD. Fatigue life of the rib-to-deck double-sided welded joints was predicted by using the mixed mode Paris law model. The influence of initial crack on the fatigue life of the outside weld toe was discussed. The major conclusions are as follows:

a. Under the vehicle load, the equivalent SIF range values of the crack fronts at the weld root of the rib-to-deck double-sided welded joints were found to be far less than the SIF threshold, and the fatigue crack did not grow.
b. The mixed mode fatigue crack growth of the weld toe was characterized by the presence of the three modes of fracture which was dominated by mode I. Mode II interacts with mode III and causes deflection of the crack surface, resulting in the final crack shape exhibiting the characteristic of a spatial curved surface.
c. The two ends of the crack fronts at the weld toe grow faster than the midpoint of the crack front due to the fact that the two ends of the crack fronts were under the action of higher values of K_{eq}. This resulted in a progressive decrease in the aspect ratio a_0/c_0 of fatigue cracks in the later phase of the crack growth.
d. Under the same fatigue load, the fatigue crack at the outside weld toe exhibited a longer fatigue life than that at the inside weld toe.
e. The initial crack exhibited a significant effect on the fatigue life of the outside weld toe of rib-to-deck double-sided welded joints of OSDs, and the flatter initial crack resulted in shortening of the fatigue life.

REFERENCES

ABAQUS. 2019. *Abaqus/CAE user's Manual*. Dassault systems, Inc.

BS 7910. 2005. Guide to methods for assessing the acceptability of flaws in metallic structures.

Chitoshi, M. 2006. Fatigue damage in orthotropic steel bridge decks and retrofit works. *Steel Structures* 6(4): 255–267.

Deng, L., Wang, W. and Yu, Y. 2016. State-of-the-art review on the causes and mechanisms of bridge collapse. *Journal of Performance of Constructed Facilities* 30(2): 04015005.

Dung, C., Sasaki, E. and Tajima, K. et al. 2014. Investigations on the effect of weld penetration on fatigue strength of rib-to-deck welded joints in orthotropic steel decks. *International Journal of Steel Structures* 15(2): 299–310.

Erdogan, F. and Sih, G. 1963 On the crack extension in plates under plane loading and transverse shear. *Journal of Basic Engineering Transactions Asme* 85(4): 519–525.

European Committee for Standardization. (CEN) 2003. EN 1991–2: Eurocode 1 Actions on structures—Part 2: traffic loads on bridges.

FRANC3D. 2016. *Franc3D Reference Manual, 7.0*. Fracture analysis consultants, Inc.

Fu, Z. Q., Ji, B. H. and Zhang, C. Y. et al. 2017. Fatigue performance of roof and U-rib weld of orthotropic steel bridge deck with different penetration rates. *Journal of Bridge Engineering* 22(6): 04017016.

Gadallah, R., Osawa, N. and Tanaka, S. 2017. Evaluation of stress intensity factor for a surface cracked butt welded joint based on real welding residual stress. *Ocean Engineering* 138: 123–139.

Hobbacher, A. 2008. *Recommendations for fatigue design of welded joints and components.* Paris, France.: International Institute of Welding.

Kainuma, S., Yang, M. and Jeong, Y. S. et al. 2016. Experiment on fatigue behavior of rib-to-deck weld root in orthotropic steel decks. *Journal of Constructional Steel Research* 119: 113–122.

Kolstein, M. H. 2007. Fatigue classification of welded joints in orthotropic steel bridge decks. *Doctoral Dissertation Delft University of Technology, Delft, Netherlands.*

Liu, Y., Chen, F. H. and Lu, N. W. et al. 2019. Fatigue performance of rib-to-deck double-side welded joints in orthotropic steel decks. *Engineering Failure Analysis* 105: 127–142.

Liu Y. Chen, F., Wang D. and Lu, N. 2020. Fatigue crack growth behavior of rib-to-deck double-sided welded joints of orthotropic steel decks. *Advances in Structural Engineering*, doi: 10.1177/1369433220961757.

Liu, Y. M., Zhang, Q. H. and Cui, C. et al. 2016. Numerical simulation method for 3D fatigue crack propagation of orthotropic steel bridge deck. *China Journal of Highway and transport* 29(7): 89–95.

Lu, N. W., Ma, Y. F. and Liu, Y. 2019. Evaluating probabilistic traffic load effects on large bridges using long-term traffic monitoring data. *Sensors* 19: 5056.

Lu, N. W., Beer, M. and Noori, M. et al. 2017a. Lifetime deflections of long-span bridges under dynamic and growing traffic load. *Journal of Bridge Engineering* 22(11): 04017086.

Lu, N. W., Noori, M. and Liu, Y. 2017b. Fatigue reliability assessment of welded steel bridge decks under stochastic truck loads via machine learning. *Journal of Bridge Engineering* 22(1): 04016105.

Lv, F., and Zhou, C. Y. and Chen, R. J. et al. 2018. A numerical analysis based on M-integral about the interaction of parallel surface cracks in an infinite plate. *Theoretical and Applied Fracture Mechanics* 96: 370–379.

Mao, J. X., Wang, H. and Li, J. 2019. Fatigue reliability assessment of a long-span cable-stayed bridge based on one-year monitoring strain data. *Journal of Bridge Engineering* 24(1): 05018015.

Masahiro, S., Naoto, N. and Akiko, T. et al. 2014. Effect of inside fillet weld on fatigue durability of orthotropic steel deck with trough ribs. *Anthology of Steel Structure Theory* 21(81): 65–77.

Nikfam, M. R., Zeinoddini, M., Aghebati, F. et al. 2019. Experimental and XFEM modelling of high cycle fatigue crack growth in steel welded T-joints. *International Journal of Mechanical Sciences* 153–154: 178–193.

Paris, P. and Erdogan, F. 1963. A critical analysis of crack propagation laws. *Journal of Basic Engineering Transactions Asme* 85(4): 528–533.

Rice, J. R. 1968. A path independent integral and the approximate analysis of strain concentration by notches and cracks. *Journal of Applied Mechanics* 35(2): 379–386.

Shi, G. Y., Li, X. X. and Zhang, G. N. 2013. Evaluation of stress intensity factor range in the prediction of fatigue crack growth at rib-to-deck welded joints of orthotropic steel decks. *Advanced Materials Research* 671–674(1): 969–973.

Song, Y. S., Ding, Y. L. and Wang, G. X. et al. 2016. Fatigue-life evaluation of a high-speed railway bridge with an orthotropic steel deck integrating multiple factors. *Journal of Performance of Constructed Facilities* 30(5): 04016036.

Wang, B., De Backer, H. and Chen, A. 2016. An XFEM based uncertainty study on crack growth in welded joints with defects. *Theoretical and Applied Fracture Mechanics* 86: 125–142.

Wang, B., Nagy, W. and De Backer, H. et al. 2019a. Fatigue process of rib-to-deck welded joints of orthotropic steel decks. *Theoretical and Applied Fracture Mechanics* 101: 113–126.

Wang, Q., Ji, B. and Fu, Z. et al. 2019b. Evaluation of crack propagation and fatigue strength of rib-to-deck welds based on effective notch stress method. *Construction and Building Materials* 201: 51–61.

Wu, W., Kolstein, M. H. and Veljkovic, M. 2019. Fatigue resistance of rib-to-deck welded joint in OSDs, analyzed by fracture mechanics. *Journal of Constructional Steel Research* 162: 105700.

Xiao, Z., Yamada, K. and Ya, S. et al. 2008. Stress analyses and fatigue evaluation of rib-to-deck joints in steel orthotropic decks. *International Journal of Fatigue* 30(8): 1387–1397.

Ya, S., Yamada, K. and Ishikawa, T. 2011. Fatigue evaluation of rib-to-deck welded joints of orthotropic steel bridge deck. *Journal of Bridge Engineering* 16(4): 492–499.

Yau, J. F., Wang, S. S. and Corten, H. T. 1980. A mixed-mode crack analysis of isotropic solids using conservation laws of elasticity. *Journal of Applied Mechanics* 47(2): 335–341.

You, R., Liu, P. and Zhang, D. et al. 2018. Experiment on fatigue performance of rib-to-deck inside welded connection in orthotropic steel decks. *Journal China and Foreign Highway* 38(3): 174–179.

Zhang, H. P., Liu, Y. and Deng, Y. 2020. Fatigue crack assessment for orthotropic steel deck based on compound Poisson process. *Journal of Bridge Engineering* 25(8), 04020057.

9 Maximum Probabilistic Traffic Load Effects on Large Bridges Based on Long-term Traffic Monitoring Data

Naiwei Lu

Changsha University of Science and Technology, China

Yafei Ma

Changsha University of Science and Technology, China

Yang Liu

Hunan University of Technology, China

CONTENTS

9.1 INTRODUCTION

Steady growth in the global transportation market has led to a dramatic increase in the highway traffic load over the past decades. In 2018, the annual growth rate of freight traffic volume in China was roughly 6%, three times that of European countries (Leahy et al. 2016; Lu et al. 2017b; Han et al. 2015). As a result, the current traffic load may exceed the design value in some design specifications that were evaluated based on traffic data a few decades ago. Traffic growth and truck overloading may result in risk sources for serviceability and safety of existing bridges. In fact, a large number of bridges all over the world have been damaged or even collapsed under heavy traffic loading (Deng et al. 2018; Deng et al. 2019). In comparison with short-span bridges, a large bridge supports higher traffic loads including larger traffic volume and the simultaneous presence of heavy trucks that govern the critical state of large bridges (Wang et al. 2014; Zhou et al. 2019). This phenomenon leads to complications in evaluating traffic load effects on long-span bridges. Hence, actual traffic patterns should be considered for probabilistic evaluation of traffic load effects on long-span bridges.

In general, structural health monitoring (SHM) systems are commonly used to investigate traffic load effects on bridges (Wang et al. 2014; Deng et al. 2018). These monitoring data can be directly utilized to evaluate various conditions of the bridge, e.g., fatigue damage, serviceability, and durability (Lu et al. 2018a; Han et al. 2019; Guo et al. 2019; Ma et al. 2019). In addition to monitoring bridge responses (stresses, displacements, and accelerations) with SHM systems installed in long-span bridges, an alternative approach is a numerical simulation with site-specific weigh-in-motion (WIM) measurements. A WIM system is usually installed with loop and piezo sensors under the road pavement to monitor traffic loads dynamically without traffic interruption (Rys 2019). With the developments in sensor technology and computational methods (He et al. 2019), the WIM system has been developed for a REMOVE strategy associated with control of trucking overloading in Europe (Jacob and Loo, 2008). In addition to traffic management, it can also provide a large amount of real traffic data for truck overloading identification and control. In practice, the WIM data has a wide range of applications in bridge engineering, such as live-load calibration in design specifications, fatigue reliability assessment, as well as lifetime maximum traffic load effects evaluation (Lu et al. 2017; Yan et al. 2017). One of these achievements is the application of static load effects in the extrapolation of extreme value by conventional methods, such as general extreme value theory and level-crossing theory (Obrien et al. 2015).

In addition to short- and medium-span bridges (Enright and O'Brien, 2013), long-span bridges have been investigated accounting for traffic loading patterns. The simultaneous presence of heavy trucks with mixtures of various statistical distributions makes the traffic pattern on long-span bridges more complicated (Caprani et al. 2016). The Monte Carlo simulation (MCS) and cellular automaton (CA) technology are popular in this regard (Wu et al. 2015). Researchers have demonstrated that congested traffic flows have a significant influence on the maximum traffic load effect on long-span bridges (Lipari et al. 2018; Xia and Ni 2016). In this respect, Xia and Ni (2016) investigated the extreme stress of Tsing Ma Bridge with a return period of 120 years using SHM data. Ruan et al. (2017) presented a site-specific traffic load model for cable-stayed bridges, which can provide a reference for similar bridges. Lu et al. (2018) investigated the lifetime deflection of a cable-stayed bridge considering multiple traffic densities based on WIM data. Micu et al. (2019) found the maximum traffic load effect on a suspension bridge based on vehicle length from image data. Yu et al. (2019) predicted the maximum bridge load effects considering traffic growth using a non-stationary Bayesian method. Although extensive studies have been conducted to evaluate the lifetime maximum load effects with WIM measurements, the application of these data for assessing the serviceability of long-span bridges is essential. In addition, most of the studies in this field have focused on the extrapolation of short-span bridges. Due to the special traffic pattern (the traffic density and traffic gap) on long-span bridges, statistical extrapolation is still a challenge. One of the difficulties in this area is how to simulate the stochastic traffic flows on long-span bridges. Moreover, the probabilistic modeling of load effects requires time-consuming computations.

This study aims to evaluate probabilistic traffic load effects on long-span bridges using traffic monitoring data. Initially, millions of WIM measurements collected from a suspension bridge in China were introduced and utilized for probability density fitting of traffic parameters. A hybrid traffic simulation method was presented by combining MCS and CA approaches to simulate multiscale critical stochastic traffic loading scenarios. A methodology for evaluating probabilistic maximum traffic load effects was developed based on level-crossing theory. A case study of the lifetime deflection of a suspension bridge was conducted to demonstrate the effectiveness of the presented traffic load model and corresponding computational framework. The daily maximum traffic load effects were estimated via the critical influence lines of the bridge. The probabilistic extrapolation was conducted based on Rice's level-crossing formula. Traffic growth and truck overloading control were considered in the parametric study to emphasize the application prospect of the present study.

9.2 TRAFFIC MONITORING DATA AND TRAFFIC FLOW SIMULATION

9.2.1 Traffic Data from WIM System

The traffic data in the present study were collected from a WIM system on a highway bridge crossing the Yangtze River located on the Yibin-Luzhou highway in Sichuan, China. The traffic data were used as probabilistic database to simulate stochastic traffic flows. Some of the onsite photos are shown in Figure 9.1. A detailed illustration of these data can be found in (Lu et al. 2019b; Lu et al. 2017a), where these data have been used for fatigue reliability evaluation of orthotropic steel decks of the bridge. Table 9.1 summarizes the general information of these data. The maximum gross vehicle weights (GVWs) of individual trucks were evaluated from the annual traffic data. The overloaded trucks were filtrated according to the Limits of Dimensions, Axle load, and Masses for road vehicles in the National Standard of the People's Republic of China (GB 1589–2016), where the threshold weights for 5-axle and 6-axle trucks are 500 and 550 kN, respectively.

It is well known that driving speed impacts the WIM sensor data due to the dynamic effect. In the present study, the vehicle data collected from the WIM system were comprehensively processed in a packed bundled software. In other words, the WIM system provided the vehicle weight data directly while considering both

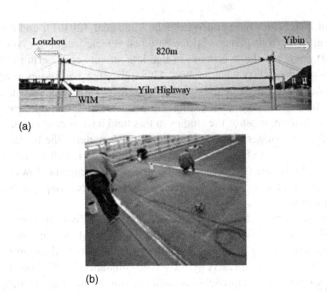

FIGURE 9.1 Onsite photos: (a) Nanxi Yangtze River Bridge; (b) weigh-in-motion (WIM) system during construction.

TABLE 9.1
Introduction of the WIM Measurements

Items	Values
Duration	Jan. 1, 2017 to Dec. 31, 2017
Recording days	365
Daily truck traffic volume	982
Traffic lanes	4
Maximum GVW (kN)	1524
Total overloaded trucks	12,521

driving and environmental effects. Since vehicle types are varied, it is difficult to consider all types of vehicles in actual configurations. This study divided vehicles into six types according to the number of axles. As types of vehicle include multi-configurations, the representative configuration with highest probability density was adopted in the present study. A detailed illustration of the vehicle classification utilized in the present study is shown in Table 9.2.

In the present study, the V6 vehicles were used as an example to demonstrate the probability distribution of vehicle weights. The probability models for GVWs and axle weights are shown in Figure 9.2a and Figure 9.2b, respectively. It is observed that the vehicle weight follows a bi-modal Gaussian distribution, which is well fitted by a Gaussian mixture model (GMM). The two peaks of the histograms correspond to the empty state and fully loaded state of trucks. With increased loading, the spacing between the two peaks will become wider. The bi-modal Gaussian distribution behavior of the truck loads are in accordance with most of the literature (Yan et al.

TABLE 9.2

Vehicle Configurations and Proportions

Vehicle Type	Description	Configuration (m)	Proportion (%)
V_1	Light truck	2.73 — AW_{11} AW_{12}	27.59
V_2	2-axle truck	5.0 — AW_{21} AW_{22}	31.23
V_3	3-axle truck	4.8 1.35 — AW_{31} $AW_{32}AW_{33}$	4.15
V_4	4-axle truck	3.75 8.6 1.31 — AW_{41} AW_{42} $AW_{43}AW_{44}$	10.44
V_5	5-axle trucks	3.6 6.8 1.31 1.31 — AW_{51} AW_{52} $AW_{53}AW_{54}AW_{55}$	10.78
V_6	6-axle truck	3.3 1.3 7.34 1.31 1.31 — AW_{61} $AW_{62}AW_{63}$ $AW_{64}AW_{65}AW_{66}$	15.82

2017; Lu et al. 2019). In addition to the GMM, alternative models, such as Lognormal mixture models and Gumbel mixture models, can also be used for the fitting.

Compared to short- or medium-span bridges, the vehicle spacing/gap has a remarkable influence on the load effect on long-span bridges. The vehicle spacing in the present study was calculated based on the driving speed and duration of the two following vehicles. Figure 9.2c and Figure 9.2d plot the statistical histogram of traffic volume and vehicle spacing in the slow lane, respectively. It is observed that busy traffic is mostly concentrated between 9:00 and 19:00, which is in line with normal working hours. Since this study focuses on the serviceability limit state of bridges, it is reasonable to select a busy traffic period for the following vehicle spacing statistical analysis.

Note that the PDFs shown in Figure 9.2 were fitted using an annual traffic data including weekdays and weekends. In practice, the traffic patterns on weekdays and weekends are different. Some researchers excluded the traffic data from weekends. For instance, OBrien et al. (2014) considered 250 working days per year to investigate the maximum traffic load effect. However, in the present study all of the traffic data were included to make the fitted PDFs more comprehensive.

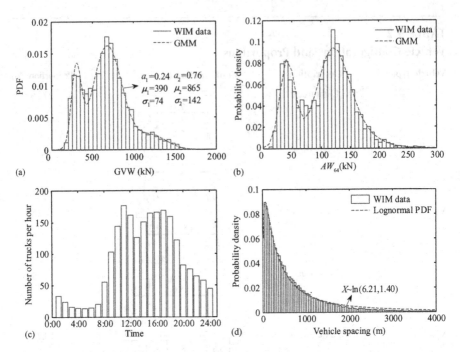

FIGURE 9.2 Probability densities of traffic parameters based on WIM data: (a) gross vehicle weight (GVW) of 6-axle trucks; (b) the fourth axle weight of the 6-axle trucks; (c) traffic density; (d) vehicle spacing in the slow lane.

9.2.2 TRAFFIC FLOW SIMULATION

With the traffic probability model, it is important to use a reasonable approach to simulate the actual traffic pattern. In general, the conventional MCS is a popular approach for simulating traffic flows (Sun and Timofeyev, 2014). However, MCS is neither an effective nor efficient tool for simulating the acceleration and deceleration of an individual vehicle. In order to address these shortcomings, numerous traffic models have been developed, on which an extensive review was conducted by Pel et al. (2012). The CA model was developed through numerous research studies to overcome the MCS shortcomings in driving behavior of individual vehicles in local level. More details regarding the application of CA model in traffic simulation can be found in Chen et al. (2013). In the present study, the CA model was used to simulate the local driving behavior of individual vehicles on the bridge.

In a CA model, the traffic space is assumed as a large number of cell grids, and each vehicle is placed in a cell. At each time step, each vehicle accelerates, decelerates, or moves depending on the predefined rules shown in Equations 9.1–9.3 (Chen and Wu 2011).

If $v_t^i < v_{max}$ and $gap_{t+1}^i \geq v_t^i + 1$, then,

$$v_{t+1}^i = v_t^i + 1 \tag{9.1}$$

If $\text{gap}_{t+1}^i \le v_t^i + 1$, then,

$$v_{t+1}^i = v_t^i - 1 \qquad (9.2)$$

Otherwise,

$$v_{t+1}^i = v_t^i \qquad (9.3)$$

where, v_t^i and v_{t+1}^i denote velocity in terms of cell/s for the ith vehicle at time t and $t + 1$, respectively; v_{max} denotes the speed limit, and gap_{t+1}^i is the gap between two vehicles in a traffic lane in terms of cell/s.

In order to capture the multiscale behavior of traffic loading on bridges, MCS and CA were combined to simulate a hybrid model of stochastic traffic load. The global-scale vehicles are simulated based on actual probability distributions as shown in Figure 9.2 via MCS, and the local-scale driving behavior of an individual vehicle is controlled by CA rules as shown in Figure 9.3. A general framework of the hybrid method is summarized in Figure 9.4.

In Figure 9.4, N_{ADT} is the average daily traffic volume. The detailed steps are out-lined as follows. Initially, the collected WIM data was filtrated to remove the invalid vehicle data, where the probabilistic traffic parameters, such as the vehicle type pro-portion, and GVW were evaluated. Second, the global parameters of individual vehi-cles including vehicle configurations, driving lane, driving speed, and GVWs were generated via MCS. Third, the vehicle spacing/gap considering dense traffic or free flowing traffic was simulated. Finally, the individual vehicle driving behavior includ-ing deceleration and changing driving lanes was determined via CA rules. Based on the above procedures, a multiscale traffic model can be simulated, where the global-scale parameters were considered in MCS, and the local-scale behavior considered in CA rules. Based on the statistical database of the WIM measurements presented above, a multiscale stochastic traffic model was simulated as shown in Figure 9.5.

Figure 9.5a shows the large scale traffic in 60-min simulated via MCS, where each marker indicates a vehicle including vehicle types, driving lanes, and GVWs. It is observed that the slow lane includes numerous overloaded trucks, especially V5 and V6 trucks. Figure 9.5b shows the dynamic driving behavior of the vehicles in 200 s in local scale coordination via CA rules. In the CA rule, the length of each cell is 5 m, the time step is 1 s, the number of cells is 200, the density of vehicles is 0.027, and the probability of changing driving lanes is 0.5. The pattern of changing driving lane is similar to overtaking. This means if a vehicle is following another vehicle, the vehicle behind has a 50% probability of changing lanes and overtaking the front vehicle if the adjacent lane is free. By extension, therefore, it also has a 50%

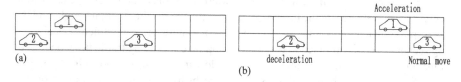

FIGURE 9.3 Cellular automaton (CA) rules: (a) $T = t$; (b)$T = t + 1$.

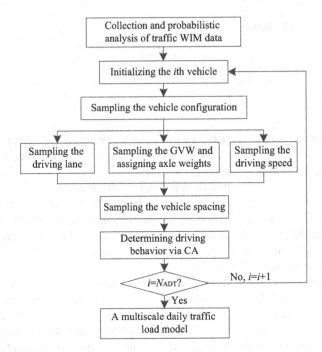

FIGURE 9.4 A hybrid traffic simulation method based on Monte Carlo simulation (MCS) and CA.

probability of following the front vehicle. This value and driving pattern is referred to by Chen and Wu (2011).

9.2.3 CRITICAL LOADING SCENARIOS

Since larger bridges usually have longer influence lines, the traffic loading pattern will impact the load effect significantly. However, it is a time-consuming process to compute numerous daily traffic load effects. In order to save the computational effort, this study searched the daily critical loading scenario which governs the maximum traffic load effect. A critical loading scenario approach corresponds to daily maxima, making it unnecessary to use the entire daily traffic data to estimate the daily load effect. However, it is reasonable to select the correct critical loading scenario from the daily traffic flow.

Based on the above assumption, the following steps were used to identify the critical traffic loading scenarios. Initially, the loading range was determined as the bridge length. Second, the traffic flow moves forward to evaluate the time-dependent total weight of the vehicles on the bridge. Subsequently, the maximum total weight and the corresponding loading scenarios were identified. Finally, these identified critical loading scenarios were combined to generate a simplified multi-day traffic flow model. An illustrative example for generating the critical loading scenario is shown in Figure 9.6.

In Figure 9.6 the bottom figures show two histories of the total weight on the bridge in 10 h computed using the stochastic traffic model. Daily maximum total

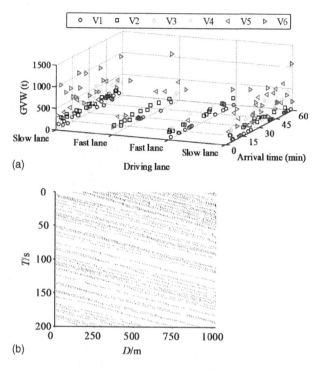

FIGURE 9.5 Traffic flow simulations: (a) global simulation via MCS; (b) local dynamic behavior of individual trucks via CA.

weights on the bridge were found in the histories, and the corresponding loading scenarios were identified in the daily traffic model as shown at the top of Figure 9.6. Finally, the identified critical loading scenarios were combined to generate a combined critical loading scenario that can be used for the following traffic load evaluation. It is worth noting that the critical loading approach identifies approximately 0.1% of the daily traffic data to evaluate the extreme load effect, which greatly reduces the time-consuming computation.

Once the critical scenarios are completely identified, these loading scenarios can be combined with sufficient spacing to generate a combined critical loading scenario as shown in Figure 9.6. With the bridge influence lines, the load effect histories can be evaluated. Subsequently, the level-crossing rate will be fitted, which will be used for extrapolation.

9.3 METHODOLOGY FOR EXTRAPOLATING MAXIMUM TRAFFIC LOAD EFFECTS

9.3.1 THEORETICAL BASIS

Rice's level-crossing principle (Chen et al. 2015), as shown in Figure 9.7, was utilized in the present study as a probabilistic basis for extrapolation. Note that Rice's formula for extrapolation is only effective for random variables following Gaussian

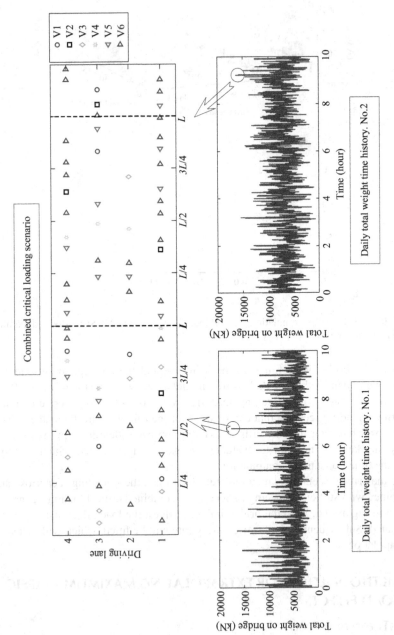

FIGURE 9.6 An example of identifying daily critical traffic loading scenarios.

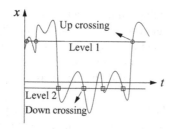

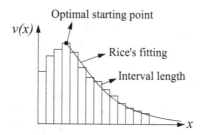

FIGURE 9.7 Rice's level-crossing principle: (a) counting number of crossings; (b) fitting level-crossing rate.

distribution (Chen 2014). In general, it can be assumed that the traffic load effect follows Gaussian distribution according to the opening literature (Caprani et al. 2016). Thus, Rice's formula was used in the present study to extrapolate traffic load effects.

As shown in Figure 9.7, the up-crossing rate is the kernel parameter to extrapolate maximum value. It is common to use $v(x)$ for a reference period and a threshold value (Cremona 2001):

$$v(x) = \frac{\sigma'}{2\pi\sigma} \exp\left[-\frac{(x-m)^2}{2\sigma^2}\right] \tag{9.4}$$

where x is the load effect which is random in nature, and m and σ are the mean value and standard deviation, respectively. When using Rice's formula for extrapolating, a cumulative distribution function (CDF) is essential which is given by

$$F(x) = \exp\left[-v_0 R_t \exp\left(-\frac{1}{2}\left(\frac{x-m}{\sigma}\right)^2\right)\right] \tag{9.5}$$

where v_0 is equal to $\sigma'/2\pi$, and R_t is the return period.

It is worth noting that the extrapolating accuracy mostly depends on the starting point and the standard derivation. Therefore, it is important to select an optimal value to start the fitting with:

$$x_{\max}(R_t) = m_{\text{opt}} + \sigma_{\text{opt}}\sqrt{2\ln(v_{0,\text{opt}} R_t)} \tag{9.6}$$

where, m_{opt}, σ_{opt} and $v_{0,opt}$ are the optimal mean value, the optimal standard deviation, and the optimal original level-crossing rate, respectively. The optimal fitting parameters can be evaluated from the Kolmogorov test with a conventional confidence level between 0.9 and 1. Detailed optimal procedure can be found in Cremona (2001), where the application on extrapolating traffic loads and load effects on multi-span bridges is demonstrated.

Based on Rice's formula for extrapolation, the first-passage probability of the random process during a period under a threshold can be evaluated by:

$$p(a,\tau) \cong 1 - \exp\left[-\int_0^\tau v(a,t)dt\right] \tag{9.7}$$

where, a is a threshold of random variable, and t is the reference period of the bridge, which is a bridge lifetime. The above formulations provide a reasonable approach to estimate maximum load effects on long-span bridges. With the consideration of these load effects, probabilistic modeling can then be carried out using the Rice formula.

9.3.2 Computational Framework

Based on the theoretical basis illustrated above, a general computational framework is presented to connect the hybrid traffic model and the probabilistic load effect analysis. The main procedures are summarized in Figure 9.8.

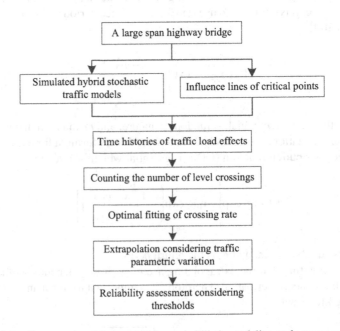

FIGURE 9.8 Computational framework for probabilistic modeling and assessment of traffic load effects.

The main procedures of the computational framework in Figure 9.8 can be simplified into three steps. First, simulate the traffic load effect histories of the bridge under the simulated traffic load model, where the traffic probability and driving behavior are included. Second, fit the optimal level-crossing model using the simulated histories. The fitting accuracy step has a significant influence on the extrapolation result. Finally, evaluate the maximum value considering a return period, and the reliability of the bridge can also be evaluated with a component resistance or threshold.

9.4 CASE STUDY

9.4.1 PROTOTYPE SUSPENSION BRIDGE

The Nanxi Yangtze River Bridge is a suspension bridge with a main span of 820 m on the Yibin-Luzhou highway located in Sichuan, China. This bridge was chosen as a prototype bridge to evaluate the maximum deflection using traffic monitoring data. The WIM measurements mentioned above were monitored for this bridge. The dimensions of the elevation layout are shown in Figure 9.9. Elaborate information of the bridge is given by Liu et al. (2015) and Lu et al. (2019). Table 9.1 indicates that a great number of overloaded trucks pass over the bridge every day. Therefore, it is an urgent task to evaluate the serviceability of the bridge under both current and future traffic loads.

9.4.2 PROBABILISTIC MODELING OF THE EXTREME LOAD EFFECTS

For the serviceability assessment of the bridge under traffic load, the girder deflection was selected for the probabilistic modeling of extreme load effects. The framework shown in Figure 9.8 was used as an outline for the computation. Initially, a finite element model was built to obtain the influence lines under a moving vehicle load. The geometric dimensions, material properties, and initial cable forces were determined by the design parameters. The influence lines of the bridge under a moving load of 100 kN were estimated in the bridge finite element model. Figure 9.10 plots the deformation influence lines of the $L/4$, 3 $L/8$, and $L/2$ points of the bridge girders. It is observed that the quarter-span is the most critical point for the maximum deflection of girders. This is in correspondence with the vertical mode shape of the bridge girders which is asymmetric. Liu et al. (2015) presented similar influence lines using the deflection monitoring data from a connected pipe system on a suspension bridge. Therefore, this study focuses on the deflection of the $L/4$ point.

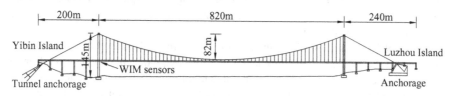

FIGURE 9.9 Dimensions of the suspension bridge.

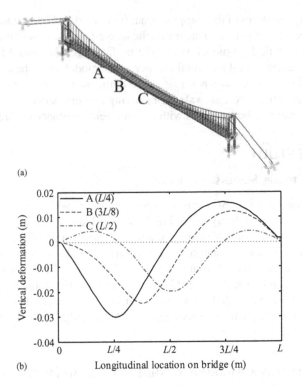

(a)

(b)

FIGURE 9.10 Critical deformation influence lines of the suspension bridge: (a) critical locations; (b) influence lines.

Deflection-time histories of the bridge were evaluated considering critical loading scenarios as given in Figure 9.6. The deflection-time history of the $L/4$ point under 10-day critical loading scenarios is shown in Figure 9.11, where $N_{-0.4}$ and $N_{-0.6}$ are the numbers of down-crossings of the threshold deflections of −0.4 m and −0.6 m, respectively. It is obvious that the number of crossings decreases with the increase of the threshold deflection. Based on 1000-day traffic loading scenarios, the histograms and fitted optimal curves were estimated as shown in Figure 9.12.

In Figure 9.12, the optimal starting point $x_{0,opt} = 0.51$ m, the optimal number of down-crossings $v_{0,opt} = 1069$, the optimal mean value $m_{opt} = -0.434$ m, the optimal standard deviation $\sigma_{opt} = 0.243$ m, and the optimal length of the intervals is 0.01 m. It is observed that Rice's fitting is close to the tail of the histograms. This indicates that the fitted curve has a good quality extrapolation. Subsequently, 1000-day block extreme values were used to investigate the deviation between generalized extreme value (GEV) results and Rice's fitting.

Figure 9.13 plots the GEV fitting and the Rice's fitting based on 1000 maxima. It is observed that while both fittings are close to the original data, they have a different trend for the extrapolations. The extrapolated maximum deflections in the 1000-year return period are −1.63 m and − 1.52 m, respectively. The deviation can be clarified in a future work by utilizing advanced fitting approaches and more data. According

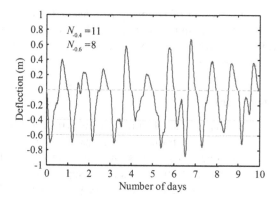

FIGURE 9.11 Time history samples of the bridge deformation under identified critical traffic loads.

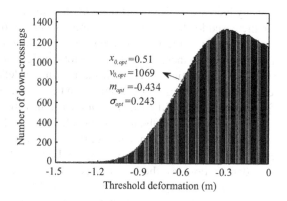

FIGURE 9.12 Histograms of Rice's level-crossing rate.

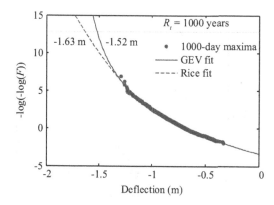

FIGURE 9.13 Comparison of Rice and generalized extreme value (GEV) fittings plotted on Gumbel probability coordinate.

to the General Code for Design of Highway Bridges and Culverts (D60–2015) in China, the threshold value should be less than the threshold $a = L/400 = 2.04$ m for the prototype bridge. Therefore, the current traffic load is far from for the serviceability limit state of the bridge.

9.4.3 PARAMETRIC STUDY

The above investigation was conducted without considering traffic growth during the bridge's lifetime. In practice, the traffic volume has a relationship with the traffic density (Richardson and Jones 2014). Thus, in order to investigate the influence of traffic growth on the traffic load effect, this study supposes that the future traffic densities are 1.2, 1.4, 1.6, 1.8, and 2.0 ρ_0, respectively. The estimated down-crossing rates are shown in Figure 9.14. It is obvious that the traffic density growth results in an increase of v_0, m, and σ in the fitting.

As observed in Figure 9.14, with an increase in traffic density the number of crossings increases and moves to the left. This phenomenon can be explained by the fact that a traffic density increase leads to a decrease in the vehicle gap on the bridge, thus resulting in an increased loading density. Therefore, the higher deflection value has a larger number of crossings. It is also observed that the peak of the curve moves to the left, which is caused by an increase in the mean total weight on the bridge.

By utilizing such a probabilistic model, both the maximum deflection and the probability of exceedance of the threshold deflection can be estimated via Equations (9.6) and (9.7). Subsequent study will focus on parametric investigation. With the level-crossing model illustrated above, the maximum deflections and the probability of exceedance in a bridge lifetime was evaluated. Initially, the maximum deflections in a 1000-year return period were estimated as shown in Figure 9.15. It is observed that when the traffic density doubles, the bridge maximum deflection in 1000-year return period increases from -1.58 to -1.85 m. In addition, the extrapolation of the deflection increases linearly with the linear growth of the traffic density.

The first-passage probabilities of the bridge under the predefined threshold were estimated accounting for traffic growth based on Equation (9.7). Figure 9.16 plots the

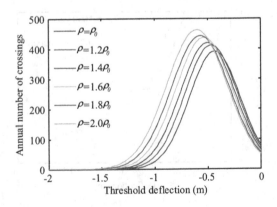

FIGURE 9.14 Influence of traffic density on the level-crossing rate.

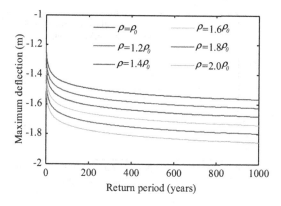

FIGURE 9.15 Maximum deflections accounting for growing traffic loads.

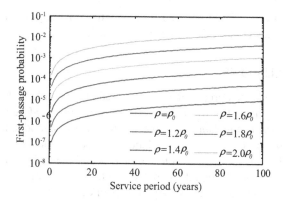

FIGURE 9.16 First-passage probability accounting for traffic growth.

probabilities of failure of the bridge in its lifetime. It is observed that when the traffic density is doubled, the first-passage probability in the bridge lifetime increases from 9.7×10^{-6} to 1.5×10^{-2}. Furthermore, the increasing rate of the first-passage probability slows down.

Since the extreme values in the time history are mostly associated with overloaded trucks, the truck overloading controls merit investigation. In China, the threshold legal weight for 6-, 5-, 4-, 3-, and 2-axle trucks is 55, 50, 40, 30, and 20 t, respectively. The stochastic traffic load model was updated, and the reliability indices of the serviceability of the bridge accounting for both traffic growth and truck overloading control are shown in Figure 9.17.

As observed from Figure 9.17, without consideration of truck overloading control, the reliability indices are 4.27 and 2.16 for the current and future traffic conditions, respectively. However, the reliability index shows a remarkable increase under the overload control. The corresponding reliability indices for current and future traffic are 6.87 and 5.08, respectively. The effective increase of the reliability index

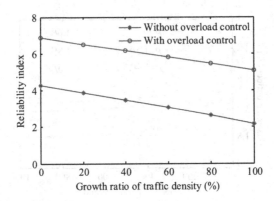

FIGURE 9.17 Reliability indices of the deformation first-passage of the bridge accounting for traffic growth and truck overloading control.

indicates that truck overloading control is essential for bridges with heavy traffic, especially under the conditions of increased traffic load.

9.5 CONCLUSIONS

This chapter investigated probabilistic traffic load effects on large bridges based on traffic monitoring data. The traffic load effects were simulated based on long-term monitored highway traffic data via a novel multiscale traffic model. It has been demonstrated that the proposed traffic simulation method has the capacity of capturing the traffic probability parameters on a global scale and individual truck driving behavior on local scale. A general computational framework was presented for maximum traffic load effect simulation. This framework and the simulated traffic model were applied to lifetime deflection evaluation of the main girder of a suspension bridge. The parametric study indicates that the traffic volume growth has a higher influence on the maximum deformation of the bridge, and thus leads to a higher probability of failure in bridge lifetime. In addition, the truck overloading control is a very efficient way of ensuring bridge reliability, and thus these controls are essential for heavy traffic areas. In terms of practical application, the numerical results can provide a theoretical basis for truck overloading control.

In addition to evaluating the maximum deformation of large bridges, the proposed traffic model and computational framework can be utilized for more aspects, such as traffic induced fatigue stresses, girder acceleration, and cable forces. However, further studies are necessary to improve the computational efficiency and accuracy. Probabilistic machine learning approaches can be used as a surrogate model to replace the time-consuming finite element runs. Consequently, a reasonable traffic growth model will be considered in the future study. Finally, parametric studies on the starting points of the fitted level-crossing model should be conducted and compared with the GEV results.

REFERENCES

Caprani, C. C. 2012. Calibration of a congestion load model for highway bridges using traffic microsimulation. *Structural Engineering International* 22: 342–348.

Caprani, C. C., OBrien, E. J. and Lipari, A. 2016. Long-span bridge traffic loading based on multi-lane traffic micro-simulation. *Engineering Structures* 115: 207–219.

Chen, S. and Wu, J. 2011. Modelling stochastic live load for long-span bridge based on microscopic traffic flow simulation. *Computers and Structures* 89: 813–824.

Chen, S. R., Nelson, R. Chen, F. and Chowdhury, A. 2013. Impact of stochastic traffic on modified cross-section profiles of a slender long-span bridge: Wind tunnel experimental investigation. *Journal of Engineering Mechanics-Asce* 139: 347–358.

Chen, W. Z., Ma, C. Xie, Z. L. Yan, B. C. and Xu, J. 2015. Improvement of extrapolation of traffic load effect on highway bridges based on Rice's theory. *International Journal of Steel Structures* 15: 527–539.

Chen, X. Z. 2014. Estimation of extreme value distribution of crosswind response of wind-excited flexible structures based on extrapolation of crossing rate. *Engineering Structures* 60: 177–188.

Cremona, C. 2001. Optimal extrapolation of traffic load effects. *Structural Safety* 23: 31–46.

Deng, Y., Li, A. Q. and Feng, D. M. 2018. Fatigue reliability assessment for orthotropic steel decks based on long-term strain monitoring. *Sensors* 18: 181.

Deng, L., Yan, W. and Nie, L. 2019. A simple corrosion fatigue design method for bridges considering the coupled corrosion-overloading effect. *Engineering Structures* 178: 309–317.

Deng, L. and Yan, W. C. 2018. Vehicle weight limits and overload permit checking considering the cumulative fatigue damage of bridges. *Journal of Bridge Engineering* 23: 04018045.

Enright, B. and O'Brien, E. J. 2013. Monte Carlo simulation of extreme traffic loading on short and medium span bridges. *Structure and Infrastructure Engineering* 9: 1267–1282.

Guo, Z., Ma, Y. Wang, L. Zhang, J. and Harik, I. 2019. Corrosion fatigue crack propagation mechanism of high strength steel bar in various environments. *Journal of Materials in Civil Engineering* 32(6), 04020115.

Han, W. S., Wu, J. Cai, C. S. and Chen, S. R. 2015. Characteristics and dynamic impact of overloaded extra heavy trucks on typical highway bridges. *Journal of Bridge Engineering* 20: 05014011.

Han, Y., Li, K. Wang, L. and Xu, G. J. 2019. Fatigue reliability assessment of long-span steel-truss suspension bridges under the combined action of random traffic and wind loads. *Journal of Bridge Engineering* 25(3), 04020003.

He, W., Ling, T. O., Brien, E. J. and Deng, L. 2019. Virtual axle method for bridge weigh-in-motion systems requiring no axle detector. *Journal of Bridge Engineering* 24: 04019086.

Jacob, B. and Loo, H. V. 2008. *Weigh-in-motion for enforcement in Europe.* In *International Conference on Heavy Vehicles HVParis 2008: Weigh-in-Motion (ICWIM 5); ISTE*: Paris, France.

Leahy, C., O'Brien, E. and O'Connor, A. 2016. The effect of traffic growth on characteristic bridge load effects. *Transportation Research Procedia* 14: 3990–3999.

Lipari, A., Caprani, C. C. and OBrien, E. J. 2018. A methodology for calculating congested traffic characteristic loading on long-span bridges using site-specific data. *Computers and Structures* 204: 65–66.

Liu, Y., Deng, Y. and Cai, C. S. 2015. Deflection monitoring and assessment for a suspension bridge using a connected pipe system: A case study in China. *Structural Control and Health Monitoring* 22: 1408–1425.

Lu, N. W., Noori, M. and Liu, Y. 2017a. First-passage probability of the deflection of a cable-stayed bridge under long-term site-specific traffic loading. *Advances in Mechanical Engineering* 9: 1–10.

Lu, N. W., Beer, M. Noori, M. and Liu, Y. 2017b. Lifetime deflections of long-span bridges under dynamic and growing traffic loads. *Journal of Bridge Engineering* 22: 04017086.

Lu, N. W., Liu, Y. and Beer, M. 2018a. Extrapolation of extreme traffic load effects on a cable-stayed bridge based on weigh-in-motion measurements. *International Journal of Reliability and Safety* 12: 69–85.

Lu, N. W., Liu, Y. and Beer, M. 2018b. System reliability evaluation of in-service cable-stayed bridges subjected to cable degradation. *Structure and Infrastructure Engineering* 14: 1486–1498.

Lu, N. W., Liu, Y. and Deng, Y. 2019a. Fatigue reliability evaluation of orthotropic steel bridge decks based on site-specific weigh-in-motion measurements. *International Journal of Steel Structures* 19: 181–192.

Lu, N. W., Noori, M. and Liu, Y. 2017c. Fatigue reliability assessment of welded steel bridge decks under stochastic truck loads via machine learning. *Journal of Bridge Engineering* 22: 04016105.

Lu, N., Ma, Y. and Liu, Y. (2019b). Evaluating probabilistic traffic load effects on large bridges using long-term traffic monitoring data. *Sensors* 19: 5056.

Ma, Y. F., Wang, G. Guo, Z. Wang, L. Jiang, T. and Zhang, J. R. 2019. Critical region method-based fatigue life prediction of notched steel wires of long-span bridges. *Construction and Building Materials* 225: 601–610.

Micu, E. A., Malekjafarian, A. OBrien, E. J. Quilligan, M. McKinstray, R. Angus, E. Lydon, M. and Catbas, F. N. 2019. Evaluation of the extreme traffic load effects on the Forth Road Bridge using image analysis of traffic data. *Advances In Engineering Software* 137: 102711.

Ministry of Communications and Transportation (MOCAT). 2015. General Code for Design of Highway Bridges and Culverts; JTG D60–2015; *Ministry of Communications and Transportation*: Beijing, China. Available online: https://www.codeofchina.com/standard/JTGD60–2015.html (accessed on December 1 2015).

National Standard of the People's Republic of China, the Limits of Dimension. 2016. *Axle Load and Masses, GB 1589*. Available online: http://www.miit.gov.cn/n1146285/n1146352/n3054355/n3057585/n3057589/c5173956/content.html (accessed on 26 July 2016).

OBrien, E. J., Bordallo-Ruiz, A. and Enright, B. 2014. Lifetime maximum load effects on short-span bridges subject to growing traffic volumes. *Structural Safety* 50: 113–122.

OBrien, E. J., Schmidt, F., Hajializadeh, D., Zhou, X. Y., Enright, B., Caprani, C. C., Wilson, S. and Sheils, E. 2015. A review of probabilistic methods of assessment of load effects in bridges. *Structural Safety* 53: 44–56.

Pel, A. J., Bliemer, M. C. J. and Hoogendoorn, S. P. 2012. A review on travel behaviour modelling in dynamic traffic simulation models for evacuations. *Transportation* 39: 97–123.

Richardson, J. and Jones, S. 2014. On the use of bridge weigh-in-motion for overweight truck enforcement. *International Journal of Heavy Vehicle Systems* 21: 83–104.

Ruan, X., Zhou, J. Y. Shi, X. F. and Caprani, C. C. 2017. A site-specific traffic load model for long-span multi-pylon cable-stayed bridges. *Structure And Infrastructure Engineering* 13: 494–504.

Rys, D. 2019. Investigation of weigh-in-motion measurement accuracy on the basis of steering axle load spectra. *Sensors* 19: 3272.

Sun, Y. and Timofeyev, I. 2014. Kinetic Monte Carlo simulations of one-dimensional and two-dimensional traffic flows: Comparison of two look-ahead rules. *Physical Review E* 89: 052810.

Wang, F. Y. and Xu, Y. L. 2019. Traffic load simulation for long-span suspension bridges. *Journal of Bridge Engineering* 24: 05019005.

Wang, W., Deng, L. and Shao, X. 2016. Number of stress cycles for fatigue design of simply-supported steel I-girder bridges considering the dynamic effect of vehicle loading. *Engineering Structures* 110: 70–78.

Wang, Y. L., Gao, Z. Y. Wang, Z. B. and Yang, J. J. 2014. A case study of traffic load for long-span suspension bridges. *Structural Engineering International* 24: 352–360.

Wu, J., Yang, F. Han, W. S., Wu, L. J., Xiao, Q. and Li, Y. W. 2015. Vehicle load effect of long-span bridges assessment with cellular automaton traffic model. *Transportation Research Record* 2481: 132–139.

Xia, Y. X. and Ni, Y. Q. 2016. Extrapolation of extreme traffic load effects on bridges based on long-term SHM data. *Smart Structures And Systems* 17: 995–1015.

Yan, D., Luo, Y. Lu, N. W. Yuan, M. and Beer, M. 2017. Fatigue stress spectra and reliability evaluation of short-to medium-span bridges under stochastic and dynamic traffic loads. *Journal of Bridge Engineering* 22: 04017102.

Yu, Y., Cai, C. S. He, W. and Peng, H. 2019. Prediction of bridge maximum load effects under growing traffic using non-stationary Bayesian method. *Engineering Structures* 185: 171–183.

Zhou, J. Y., Ruan, X. Shi, X. F. and Caprani, C. C. 2019. An efficient approach for traffic load modelling of long span bridges. *Structure and Infrastructure Engineering* 15: 569–581.

10 Dynamic Reliability of Cable-supported Bridges Under Moving Stochastic Traffic Loads

Yang Liu

Hunan University of Technology, China

Qinyong Wang

Changsha University of Science and Technology, China

Naiwei Lu

Changsha University of Science and Technology, China

CONTENTS

10.1 INTRODUCTION

In recent decades, the transportation industry has grown rapidly with the development of global economy. As a result, truck overloading and traffic volume growth have become a safety hazard for existing bridges. Large investigations (Han et al. 2015; Lu et al. 2017) show that the traffic load models in bridge design specifications may under-estimate the current traffic load effect. On the other hand, a poor pavement roughness condition due to truck overloading amplifies the traffic load effects (Zhao et al. 2017). Therefore, the actual traffic load effects might exceed the threshold or design value. Cable-stayed bridges are critical elements of a highway and have the unique characteristic of supporting simulations presence multiple trucks compared to short-span bridges. Therefore, evaluating the maximum dynamic traffic load effects on cable-stayed bridges using actual traffic loads is an urgent task. Traffic control measures can be subsequently conducted based on the numerical results to ensure bridge safety.

Since actual traffic loads are site specific and random in nature, site-specific traffic data and probabilistic approaches are commonly used for evaluating maximum traffic load effects on bridges. A vast number of WIM records were collected from five European sites by OBrien and Enright (2013) to simulate the maximum effect on short- to medium-span bridges considering free-flowing traffic flows. WIM data in Hong Kong was utilized to develop a lane load model which was then compared with several existing bridge live-load models (Miao and Chan 2002). In the Netherlands WIM measurements were used to extrapolate the maximum traffic load effect on a short-span bridge (O'Connor and O'Brien 2005) was developed to evaluate maximum traffic load effects on a multi-pylon cable-stayed bridge (Ruan et al. 2017). These studies have made great contributions to the extrapolation of traffic load effects on long-span bridges based on WIM measurements.

The conventional approach for probabilistic extrapolation is the block maxima approach, including the peaks over threshold approach and the generalized extreme value (GEV) approach (Obrien et al. 2015). These conventional approaches are effective for short-span bridges with discontinuous traffic load effects (Fu and You 2009), but not efficient for long-span bridges with a simultaneous presence of multiple trucks. The traffic density on a long-span bridge is sensitive to the maximum traffic load effect (Zhou and Chen 2016). In this regard, Rice's level-crossing theory can be used for probabilistic modeling of extreme traffic load effects. It is appropriate to count the number of crossings for long-span bridges because traffic load effect histories are mostly continuous and thus this approach can capture more information about the maximum traffic load effect than the GEV distribution approach.

The vehicle-bridge interaction is a major factor leading to the dynamic effect of traffic loads, especially for bridges with larger stiffness girders. From a probabilistic point of view, the dynamic behavior of cable-supported bridges may influence the probabilistic extreme load effect under stochastic traffic loads. The current design codes (European Committee for Standardization 2003; AASHTO 2015) and many researchers (Zhu et al. 2018; Taghinezhadbilondy et al. 2018) in this area conventionally use the dynamic amplification factor (DAF) and the dynamic load allowance (DLA) to consider the impact effects on the VBI interaction system in a critical scenario. In addition, probabilistic approaches have also been developed for describing the DAF in a probability domain. For instance, a cumulative distribution function of the DAF was developed (Fedorova and Sivaselvan 2017; Torres et al. 2019). An assessment dynamic ratio (ADR) was developed to define the relationship between the characteristic dynamic traffic load effect and the characteristic static traffic load effect in a reference period (Guo and Caprani 2019). Subsequently, the ADR was utilized as a lifetime dynamic allowance value to investigate the dynamic impact of traffic loading on short- to medium-span concrete bridges (Carey et al. 2017). The

Gumbel distribution was demonstrated (Yu et al. 2019) to be the best fit to the probability model of DAFs of a wide range of bridges. The multivariate extreme value theory and the ADR were developed to investigate the dynamic allowance of highway bridges under long-term traffic loading (Zhou et al. 2018). These studies illustrated above mostly emphasized the probabilistic dynamic effects of the VBI on short- to medium-span bridges, with less attention paid to cable-supported bridges. One of the problems is how to consider the traffic loading scenarios on long-span bridges. Integrating the traffic-bridge analysis and probabilistic modeling for probabilistic extrapolation needs further investigation.

This study investigated the dynamic extrapolation of traffic load effects on long-span cable-supported bridges using site-specific traffic data. The dynamic traffic load effect is considered in a traffic-bridge interaction system. The probabilistic maximum traffic load effects are fitted by a level-crossing formula. A computational framework is presented to illustrate the connection between the dynamic and probabilistic analyses. Two long-span cable-stayed bridges were selected as prototypes to investigate the influence of the dynamic effect on the extrapolated bridge deflection and the corresponding probability of exceedance. The influence of traffic growth on the probability of exceedance was also investigated.

10.2 THEORETICAL BASIS OF RICE'S LEVEL-CROSSING RATE

As mentioned in the introduction, the conventional approach to simulate extreme traffic load effects is associated with the GEV approach. However, the accuracy of the approach mostly depends on the number of blocks. Obviously, the GEV approach ignores most of the data that is simulated in a time-consuming computation for long-span bridges. Rice's level-crossing theory (Rice 1944) based on number of crossings is shown in Figures 10.1 (a, b) for probabilistic modeling. It is obvious that Rice's level-crossing approach can capture more information of the traffic load effect.

A precise means of implementing the level-crossing theory is the Gaussian and stationary assumption of the traffic load effect process. In general, the traffic load effect on long-span bridges can be assumed as a stationary Gaussian process or even a white noise, because the structural influence line is quite a bit longer than the length of a normal vehicle (Wang and Xu 2019). On the basis of Rice's level-crossing

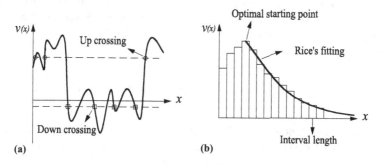

FIGURE 10.1   Principle of Rice's level-crossing theory: (a) crossings; (b) fitting to crossings.

theory, the level-crossing rate under the condition of a threshold and a reference period can be written as (Cremona 2001):

$$v(x) = \frac{\sigma'}{2\pi\sigma} \exp\left[-\frac{(x-m)^2}{2\sigma^2}\right] \tag{10.1}$$

where, x is a random process, m and σ are the mean value and standard deviation of x, respectively, and σ' is the standard deviation of the derivative of x.

In practice, $v(x)$ can be fitted to the histograms of the number of crossings, as shown in Figure 10.1(b). By dividing the entire nonstationary process into several stationary interval processes, the level-crossing rate of original process is supposed as a superposition of interval processes in proportions. Based on this assumption, the kernel coefficients can be evaluated by a second-order polynomial function (Lu et al. 2019b):

$$\ln\left[\bar{v}(x)\right] = \ln\left[\sum_{i=1}^{m} \frac{p_i N_i(x)}{T}\right] = a_0 + a_1 x + a_2 x^2 \tag{10.2}$$

where $\bar{v}(x)$ is an estimated equivalent level-crossing rate of the nonstationary process, $a_0 = \ln(v_0) - \frac{m^2}{2\sigma^2}$, $a_1 = \frac{m}{\sigma^2}$, and $a_2 = \frac{1}{-2\sigma^2}$ are second-order polynomial coefficients that can be evaluated based on the histograms of the number of crossings.

An optimal starting point and the number of intervals are critical for an accurate extrapolation. A starting point close to the tail is better for extrapolation, but worse for extrapolation. Based on the optimal starting point and corresponding number of intervals, the maximum value in a return period can be written as

$$x_{\max}(R_t) = m_{opt} + \sigma_{opt}\sqrt{2\ln(v_{0,opt}R_t)} \tag{10.3}$$

where R_t is a return period, $x_{\max}$ is the maximum value corresponding to the return period, m_{opt}, σ_{opt} and $v_{0,opt}$, represent the optimal mean value, optimal standard derivation, and optimal crossing rate, respectively.

In addition to extrapolation, the probabilistic model can also be used to estimate failure probability. First-passage failure is the best description of a stochastic process crossing the prescribed threshold during an interval time. Based on Rice's level-crossing theory and the assumption of the Poisson distribution, the probability of failure can be written as

$$P_f(a, T_s) = A\exp\left[-\int_0^{T_s} v_i(a)\,dt\right] \cong \exp\left[-T_s\bar{v}(a)\right] \tag{10.4}$$

where A is a coefficient associated with the stationarity of a random process and can be assumed as 1 for a stationary process, T_s is usually the lifetime of a bridge, and a is the threshold value for a bridge, such as $L/500$ for the deflection limit of a concrete cable-stayed bridge (JTG/T D65–01, 2007).

10.3 A COMPUTATIONAL FRAMEWORK FOR EXTRAPOLATION

It is acknowledged that the VBI leads to a fluctuation in bridge responses. On one hand, this fluctuation amplifies the peak value of the static response, while on the other, it increases the number of level crossings. Therefore, the traffic-bridge interaction will affect Rice's level-crossing model. In order to consider this phenomenon, this study presents a computational framework integrating the traffic-bridge interaction and the level-crossing theory to extrapolate the maximum traffic load effect of long-span bridges. The flow chart of the proposed computational framework is shown in Figure 10.2.

As depicted in Figure 10.2, the framework begins with the traffic load models and the bridge modal parameters. With the critical traffic loading scenarios, the bridge dynamic responses can be computed using the equivalent dynamic wheel load (EDWL) approach with consideration of the road roughness condition (RRC). Since the modal superposition approach is involved in the EDWL approach, the structural nonlinearities cannot be considered in the proposed approach. Instead of a constant daily maximum, the time histories of the dynamic analysis can provide more statistical information for modeling the level-crossing rate. The dynamic history provides a way to account the number of crossings due to the VBI.

In general, a VBI system is conventionally modeled in two subsystems including the vehicle system with the degrees of freedom (DOFs) in physical coordinates and the bridge system with the DOFs in mode coordinates of the concerned mode shapes. For long-span bridges, the simultaneous presence of multiple vehicles is a unique phenomenon different to short-span bridges. The traffic-bridge interaction (TBI) is more complex than a single VBI model because each vehicle will impact the equation of motion (Zhou and Chen 2014). An EDWL approach (Chen and Cai 2007) utilizes several moving time-varying loads evaluated from a fully coupled vibration analysis of a VBI system to represent the actual moving vehicles on a bridge.

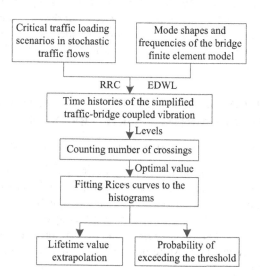

FIGURE 10.2 Computational framework for extrapolating maximum traffic load effects considering traffic-bridge interaction.

A dynamic load ratio describing the relationship between the EDWL and the weight of a vehicle is a critical factor to utilize the EDWL approach. The dynamic load ratio is associated with the physical parameters of the vehicle and the bridge dynamics. Based on the above illustration, the fully coupled TBI system can be written as

$$\mathbf{M}_b\ddot{\mathbf{u}}_b + \mathbf{C}_b\dot{\mathbf{u}}_b + \mathbf{K}_b\mathbf{u}_b = \mathbf{F}_{eq}^{wheel} \tag{10.5}$$

$$F_{eq}^{wheel}(t) = \sum_{j=1}^{n_v}\left\{\left[1 - R_j(t)\right]G_j \cdot \sum_{k=1}^{n_b}\left\{h_k\left[x_j(t) + \alpha_k\left[x_j(t)d_j(t)\right]\right]\right\}\right\} \tag{10.6}$$

where $\mathbf{F}_{eq}^{wheel}$ is the time-varying force vector on the bridge; R_j and G_j are the dynamic load ratio and the gross vehicle weight of the jth vehicle; x_j, and d_j are the longitudinal and transverse location of the gravity center of the jth vehicle, respectively; h_k and a_k are the vertical and the torsional mode shapes for the kth mode of the bridge model, respectively; and n_v and n_b are the number of vehicles and the number of adopted bridge modes, respectively. It is worthwhile to note that the road surface roughness condition is considered in the relative displacement. Note that the total number of vehicles changes with time depending on the density of the stochastic traffic flow. Nevertheless, it can be assumed to be stationary for dense traffic flow and will be considered in the present study. With the EDWLs, the dynamic load effect of the bridge under traffic loads can be efficiently calculated.

Eventually, the level-crossing rate can be fitted to the histograms of the number of level crossings. Given a return period, the maximum traffic load effect can be estimated based on Equation (10.3). In addition to the characteristic traffic load effect extrapolation, the probability of exceedance of the limit value can be evaluated from the CDF of the maximum traffic load effect as shown in Equation (10.4). Note that the efficiency of the method depends on the number of days the number of crossings are counted. In general, 1000-day simulations are adequate for fitting an effective probabilistic extreme value model. The highlight of the computational framework is the integration of the VBI analysis and the Rice level-crossing evaluation.

10.4   NUMERICAL SIMULATION OF DYNAMIC TRAFFIC LOAD EFFECTS ON CABLE-SUPPORTED BRIDGES

10.4.1   STOCHASTIC TRAFFIC LOAD SIMULATION BASED ON WIM MEASUREMENTS

In general, a WIM system is utilized for monitoring passing vehicles using the sensors installed under the bridge pavement. When a moving vehicle passes over the WIM system, the axle weight can be directly measured in the system, and the driving speed, vehicle type, and vehicle spacing can be indirectly evaluated with the recorded data time. The WIM system in the present study was referred to a suspension bridge on the Yilu highway in China. More information of the WIM data has been elaborated on in Lu et al. (2019).

One-year data in 2018 of the system were utilized for the following probabilistic modeling of traffic parameters. First of all, these data were classified into six types

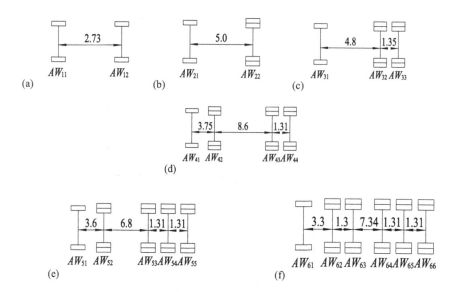

FIGURE 10.3 Vehicle configurations: (a) V1; (b) V2; (c) V3; (d) V4; (e) V5; (f) V6.

according to the vehicle configurations. The six types of vehicle configurations are shown in Figure 10.3, where V1 denotes light cars, V2–V6 denote trucks with 2 to 6 axles, AW_{ij} denotes the axle weight of the jth axle and the ith types of the truck.

Subsequently, the probability density functions (PDFs) of the vehicle weights of each type of vehicle were fitted with the Gaussian mixture model (GMM). Figures 10.4 (a, b) show the axle weight and the gross vehicle weight (GVW) of the V6 vehicles, where a, μ, and σ represent the weight, mean value, and standard deviation of the weight. It is observed that the axle weight and the GVW have multiple peaks due to truck overloading, and the GMM fits the histograms fairly well.

The vehicle spacing between two following vehicles is another parameter impacting the traffic load effect on bridges. In addition, most of the extreme value is generated by dense or congested traffic. Therefore, statistical analysis of the vehicle spacing was conducted using the driving speed and recording time data. In order to consider the traffic density on the bridge, all of the vehicle spacings were divided into dense traffic and free traffic according to a threshold value of a time interval of 2 s (Lu et al. 2017a). The two types of PDFs of the vehicle spacing are shown in Figures 10.5 (a, b). It is observed that the vehicle spacing in free traffic follows a lognormal distribution while denser traffic follows Gamma distribution as shown in Figure 10.5 (a). Since the occupancy of dense traffic data is 2.15% in the entire data, only a small amount of the data leads to the deviation fitting of the PDF curve as shown in Figure 10.5 (b).

Based on the statistics of traffic parameters, stochastic traffic flows were simulated using the Monte Carlo simulation method. Figures 10.6 (a, b) provide the simulated stochastic traffic flows with the parameters of the vehicle type, the GVW, the driving lane, and the arrival time. In Figures 10.6 (a, b), each dot represents an individual vehicle.

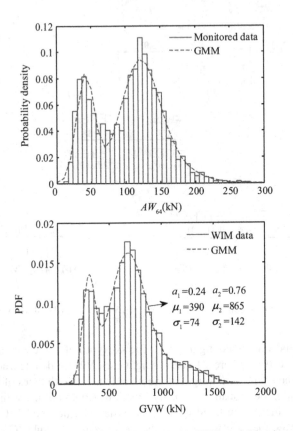

FIGURE 10.4 Histograms and PDFs of the 6-axle: (a) GVW; and (b) axle weight.

It is noted that the stochastic traffic model can be resampled to generate thousands of traffic samples, which can be utilized for subsequent probabilistic modeling of traffic load effects. For extreme traffic load analysis in the present study, the daily stochastic traffic model can be simplified into several critical loading scenarios, which will be more efficient for a large number of probabilistic simulations.

10.4.2 DYNAMIC TRAFFIC LOAD EFFECT ON PROTOTYPE BRIDGES

Two long-span cable-supported bridges were chosen as prototypes to demonstrate the application of the proposed computational framework. A cable-stayed bridge has 34 pairs of stay cables in a fan pattern for each pylon. A suspension bridge has 64 steel box-girders with a segmental length of 12.8 m in the mid-span. More details of the two bridges can be found in Lu et al. (2019). The first-order mode shapes of the two bridges in the finite element models are shown in Figure 10.7. It is observed that the first-order mode shape of the cable-stayed bridge is symmetric, while that of the suspension bridge is asymmetric.

The RRC in this study was defined by the RRC coefficient recommended by the International Organization for Standardization (ISO 1995). The RRC coefficient for

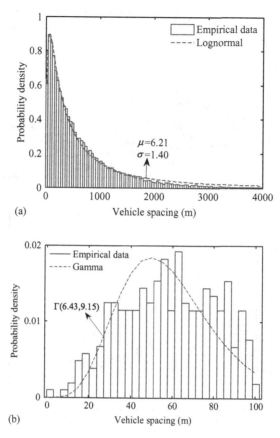

FIGURE 10.5   Histograms and PDFs: (a) free traffic; (b) dense traffic.

classifications of "Good" with an RRC of 128×10^{-6} was used in this study. Considering the dynamic characteristics of the bridges and vehicles, 1-hour dense traffic loads were simulated passing on the bridges. Figures 10.8 (a, b) plot the deflection histories of the girder points $L/2$ and $L/4$.

As observed from Figures 10.8 (a, b), the mid-span ($L/2$) point of the cable-stayed bridge has a severe vibration in comparison with the quarter-span ($L/4$) point. However, the $L/4$ point of the suspension bridge has a more severe vibration than the $L/2$ point. In addition, the suspension bridge has relatively high displacements compared to the cable-stayed bridge. The first phenomenon can be explained by the vertical first-order fundamental mode shapes of the two types of bridge as shown in Figures 10.8 (a, b), where the cable-stayed bridge is symmetric, but the suspension bridge is asymmetric. The second phenomenon is associated with the stiffness of the structures. In order to determine the difference of the deformations of the two bridges in a probabilistic domain, the mean values and root-mean-square (RMS) deflections were computed, as shown in Figures 10.9 (a, b).

It is observed from Figures 10.9 (a, b) that the RMS of the cable-stayed bridge follows a C-type distribution along the girder, while the RMS of the suspension

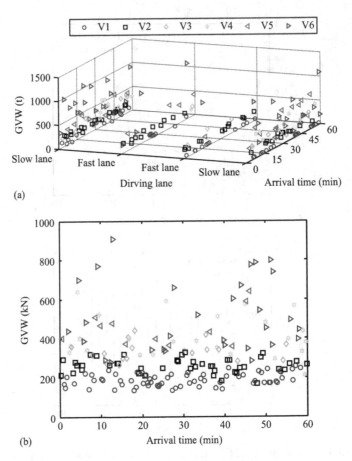

(a)

(b)

FIGURE 10.6   Stochastic traffic flow models: (a) 3D; (b) 2D.

bridge follows an M-type distribution. There is no doubt that the critical point for the cable-stayed bridge is the $L/2$ point. However, this conclusion is not the same for the suspension bridge, because the maximum RSM value occurs at $L/4$ point of the girders. In addition, the RSM value has a greater influence on the probabilistic extrapolation. The critical $L/4$ point for the suspension bridge has also been highlighted by Liu et al. (2015) using a connected pipe system. Therefore, the deflection-critical point for the cable-stayed bridge is L/2 point, and that for the suspension bridge is the $L/4$ point. The following investigations focus on the two critical points.

Following the computational framework shown in Figure 10.2, the primary step is to evaluate the time histories of the bridge girder ($L/2$ point or $L/4$ point) under critical traffic loading. Both static and dynamic traffic load effects were evaluated as shown in Figure 10.10. In Figure 10.10, a reference line is shown to count the number of level crossings, where $N_{s,i}$ and $N_{d,I}$ are the number of crossings of the static and dynamic deflections, respectively.

As observed from Figures 10.10 (a, b), the dynamic deflection history has more level crossings compared to the static deflection history. Since the cable-stayed

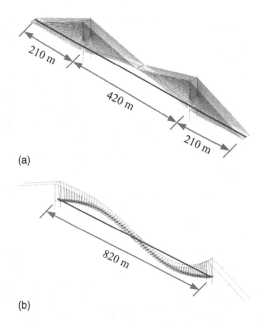

(a)

(b)

FIGURE 10.7 First-order vertical mode shapes: (a) cable-stayed bridge; (b) suspension bridge.

bridge has more stiffness than the suspension bridge, the VBI effect is more evident than for the suspension bridge. Therefore, the difference between $N_{s,i}$ and $N_{d,i}$ for the cable-stayed bridge is larger than that of the suspension bridge.

10.5 PROBABILISTIC ESTIMATION USING RICE'S FORMULA

10.5.1 MAXIMUM DEFLECTION EXTRAPOLATION

A total number of 1000-day critical loading scenarios selected from the stochastic traffic flows were utilized to count the number of level crossings. The cumulative numbers of crossings were plotted in histograms, and the Rice curves were fitted to the histograms as shown in Figure 10.11. It is observed that the cable-stayed bridge has smaller values of the mean value and the standard derivation, but a larger crossing rate.

With the normalized level-crossing data, the maximum traffic load effects were extrapolated based on Rice's formula. Given a return period of a 1000-year period, the maximum traffic load effects were extrapolated as shown in Figures 10.12 (a, b), where x_{code} values are the bridge deflections evaluated considering design traffic load models. It can be seen that the maximum deflections for the critical points of the cable-stayed bridge are 0.655 m and 0.694 m, respectively, and 1.891 m and 1.960 m, respectively, for the suspension bridge. Therefore, the DAF for the cable-stayed bridge and the suspension bridge is 5.9% and 3.6%, respectively. It can be concluded that the dynamic effect of traffic loading on long-span bridges is associated with structural stiffness, where a higher stiffness leads to a larger amplification factor.

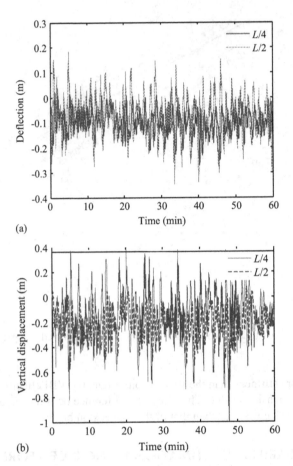

(a)

(b)

FIGURE 10.8 Deflection histories of the critical points in one hour: (a) cable-stayed bridge; (b) suspension bridge.

10.5.2 Probability of Exceedance of Threshold

A bridge deflection limitation for traffic load effects is a condition to control the undesirable vibrations of a bridge without causing unrecoverable damage. Most design specifications have defined the limit deflection of multiple bridges. The thresholds of traffic-induced bridge displacement in the design specifications of China (JTG/T D65–15 2015) are summarized in Table 10.1, where L is the bridge span length. In the present study, the materials for the girders of the cable-stayed bridge and the suspension bridge are concrete and steel, respectively. Therefore, the deflection limits for the cable-stayed bridge and the suspension bridge are $L/500 = 0.84$ m and $L/300 = 2.34$ m, respectively.

The probabilities of exceedance of the two long-span bridges were evaluated based on the probabilistic model and the deflection threshold. Figure 10.13 plots the probability of exceedance versus the bridge service period. It can be seen that the suspension bridge has a higher probability of exceedance than the cable-stayed

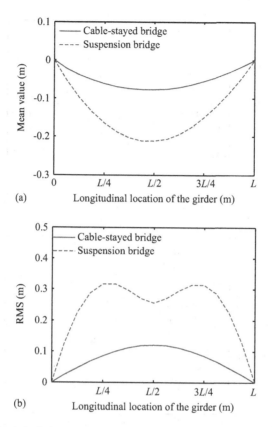

FIGURE 10.9 Statistical characteristics of the two cable-supported bridges: (a) mean value; (b) standard deviation.

bridge, even though the deformation threshold for cable-stayed bridge is higher than the suspension bridge as mentioned above.

As a parametric study, the influence of average truck traffic (ADTT) volume on the probability of exceedance was investigated. In general, ADTT is influenced by several factors that are difficult to predict in a bridge lifetime. This study assumes the annual linear growth rate to be 0%, 1%, 2%, 3%, and 4%. This hypothesis adopts the assumption that the ADTT has a jump in growth to the objective value rather than a continuous growth. Subsequently, the traffic growth was considered by decreasing the vehicle spacing in the stochastic traffic model. Influences of the growth rate of ADTT on the probability of exceedance of the bridge threshold deflection are shown in Figure 10.14.

It is observed that the probability of exceedance has a higher decrease rate at the initial stage of traffic growth, and this increase rate slows down with the continued growth of ADTT. In addition, the growth rate of ADTT widens the gap between the two values of the probability of failure. This has demonstrated that a more realistic traffic growth model is still important for bridge reliability assessment.

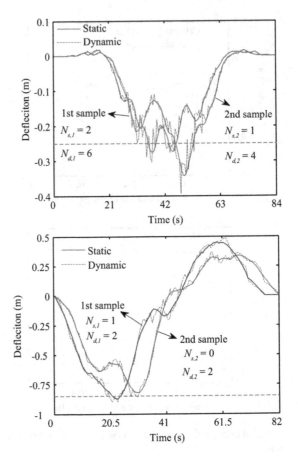

FIGURE 10.10 Deflection histories: (a) **L/2** point of the cable-stayed bridge; (b) **L/4** point of the suspension bridge.

10.6 CONCLUSIONS

A probabilistic method was presented to evaluate the dynamic traffic load effects on cable-supported bridges. The weigh-in-motion measurements of a highway bridge in China were utilized to simulate the actual traffic load on bridges. Both dynamic and probabilistic traffic load effects were considered in the methodology. Rice's extrapolation theory was utilized to capture the impact of the TBI on the probabilistic extreme traffic load effect. Case studies of a cable-stayed bridge and a suspension bridge were conducted to demonstrate the feasibility of the proposed methodology. The conclusions are summarized as follows:

1. Rice's level-crossing theory can capture the probabilistic characteristics of the TBI effect, and thus provides a reliable extrapolation with the consideration of dynamic factors, e.g., road surface roughness conditions, bridge stiffness, and vehicle suspension system.

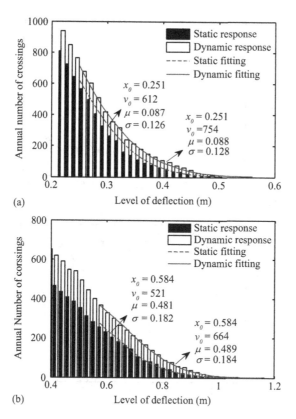

FIGURE 10.11 Histograms and Rice's fittings of the number of level crossings: (a) cable-stayed bridge; (b) suspension bridge.

2. The RMS displacement of the cable-stayed bridge follows a C-type distribution, while that for the suspension bridge follows an M-type distribution. This distribution is associated with the first-order mode shapes of the two types of bridges. The deflection-critical points for the cable-stayed bridge and suspension bridge under traffic loads are the mid- and quarter-span points, respectively.

3. The DAF decreases with the increase of bridge length or the decrease of bridge stiffness. The amplification factors for the cable-stayed bridge with a mid-span length of 420 m and the suspension bridge with a mid-span length of 820 m are 5.9% and 3.6%, respectively. Therefore, the DAF for extrapolation is negligible for a flexible long-span bridge.

4. The traffic growth ratio has a higher influence on the probability of exceedance of the threshold deflection for the suspension bridge. A linear annual growth ratio of 4% results in the probability of exceedance increasing from 4.0×10^{-5} to 2.1×10^{-2}. The traffic growth is more significant for the extrapolation of traffic load effects on long-span bridges.

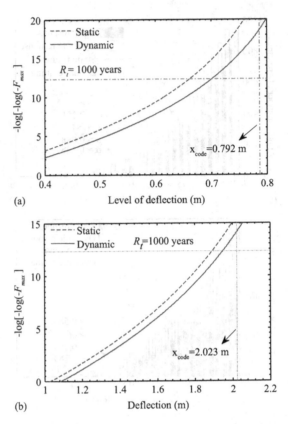

FIGURE 10.12 Extrapolation of the extreme traffic load deflections: (a) cable-stayed bridge; (b) suspension bridge.

TABLE 10.1
Threshold Deflection of Different Types of Bridges

Bridge types	Materials of bridge girders	Deflection limits
Simply supported bridges	Concrete or steel	$L/800$
Cantilever bridge	Concrete or steel	$L/300$
Cable-stayed bridges	Concrete	$L/500$
	Steel	$L/400$
Suspension bridges	Steel	$L/350$

In addition to the application for cable-supported bridges, the proposed computational framework can also be applied to any other type of bridge. However, some challenges remain to be addressed in future works. First, the time-consuming TBI analysis is still a bottleneck for the computational efficiency of the computation framework. Second, the probability model and extrapolation of the extreme traffic load effect will be compared with other approaches, such as the GEV distribution and

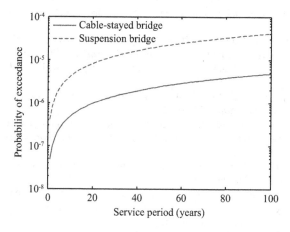

FIGURE 10.13 Probability of exceedance to the deflection limit.

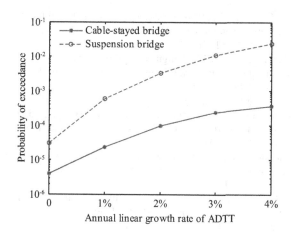

FIGURE 10.14 Influence of growth rate of ADTT on the probability of exceedance.

the peaks over value approach. Finally, a more realistic traffic growth model will be considered instead of an idealized traffic growth model in a further study.

REFERENCES

AASHTO. (2015). *LRFD bridge design specifications*, 6th Ed., Washington, DC.
European Committee for Standardization, 2003, *Actions on structures. Part 2: traffic loads on Bridges*, Brussels,
Carey, C., OBrien, E. J. Malekjafarian, A. Lydon, M. and Taylor, S. 2017. Direct field measurement of the dynamic amplification in a bridge. *Mechanical Systems and Signal Processing* 85: 601–609.
Chen, S. R. and Cai, C. S. 2007. Equivalent wheel load approach for slender cable-stayed bridge fatigue assessment under traffic and wind: Feasibility study. *Journal of Bridge Engineering* 12(6): 755–764.

Cremona, C. 2001. Optimal extrapolation of traffic load effects. *Structural Safety* 23(1): 31–46.

Fedorova, M. and Sivaselvan, M. V. 2017. An algorithm for dynamic vehicle-track-structure interaction analysis for high-speed trains. *Engineering Structures* 148: 857–877.

Fu, G. and You, J. 2009. Truck loads and bridge capacity evaluation in China. *Journal of Bridge Engineering* 14(5): 327–335.

Guo, D. and Caprani, C. C. 2019. Traffic load patterning on long span bridges: A rational approach. *Structural Safety* 77: 18–29.

Han, W., Wu, J. Cai, C. S. and Chen, S. 2015. Characteristics and dynamic impact of over-loaded extra heavy trucks on typical highway bridges. *Journal of Bridge Engineering* 20(2): 05014011.

ISO (1995), *Mechanical vibration-road surface profiles-reporting of measured data, ISO 8069: E*, Geneva, Switzerland.

Liu, Y., Deng, Y. and Cai, C. S. 2015. Deflection monitoring and assessment for a suspension bridge using a connected pipe system: A case study in China. *Structural Control and Health Monitoring* 22(12): 1408–1425.

Lu, N. W., Beer, M. Noori, M. and Liu, Y. 2017a. Lifetime deflections of long-span bridges under dynamic and growing traffic loads. *Journal of Bridge Engineering* 22(11): 04017086.

Lu, N. W., Noori, M. and Liu, Y. 2017b. First-passage probability of the deflection of a cable-stayed bridge under long-term site-specific traffic loading. *Advances in Mechanical Engineering* 9(1): 1687814016687271.

Lu, N. W., Liu, Y. and Deng, Y. 2019a. Fatigue reliability evaluation of orthotropic steel bridge decks based on site-specific weigh-in-motion measurements. *International Journal of Steel Structures* 19(1): 181–192.

Lu, N. W., Ma, Y. F. and Liu, Y. 2019b. Evaluating probabilistic traffic load effects on large bridges using long-term traffic monitoring data. *Sensors* 19(22): 5056.

Miao, T. J. and Chan, T. H. 2002. Bridge live load models from WIM data. *Engineering Structures* 24(8): 1071–1084.

OBrien, E. J., Schmidt, F. Hajializadeh, D. Zhou, X. Y. Enright, B. Caprani, C. C. and Sheils, E. 2015. A review of probabilistic methods of assessment of load effects in bridges. *Structural Safety* 53: 44–56.

OBrien, E. J. and Enright, B. 2013. Using weigh-in-motion data to determine aggressiveness of traffic for bridge loading. *Journal of Bridge Engineering* 18(3): 232–239.

O'Connor, A. and O'Brien, E. J. 2005. Traffic load modelling and factors influencing the accuracy of predicted extremes. *Canadian Journal of Civil Engineering* 32(1): 270–278.

Rice, S. O. 1944. Mathematical analysis of random noise. *Bell System Technical Journal* 23(3): 282–332.

Ruan, X., Zhou, J. Shi, X. and Caprani, C. C. 2017. A site-specific traffic load model for long-span multi-pylon cable-stayed bridges. *Structure and Infrastructure Engineering* 13(4): 494–504.

Specifications for design of Highway Cable-stayed Bridge, *JTG/T D65-01-2007*, (Beijing, China Communications Press, Approved January 2007). (http://www.gbstandards.org/GB_standard_english.asp?code=JTG/T%20D65-01-2007&word=Guidelines%20for%20Design%20of%20Highw).

Specifications for design of Highway Suspension Bridge, *JTG/T D65–15* (Beijing, China Communications Press, approved March 2015). (http://www.gbstandards.org/index/GB_standard_english.asp?id=74,890&word=Specifications%20for%20D).

Taghinezhadbilondy, R., Yakel, A. and Azizinamini, A. 2018. Deck-pier connection detail for the simple for dead load and continuous for live load bridge system in seismic regions. *Engineering Structures* 173: 76–88.

Torres, V., Zolghadri, N. Maguire, M. Barr, P. and Halling, M. 2019. Experimental and analytical investigation of live-load distribution factors for double tee bridges. *Journal of Performance of Constructed Facilities* 33(1): 04018107.

Wang, F. Y. and Xu, Y. L. 2019. Traffic load simulation for long-span suspension bridges. *Journal of Bridge Engineering* 24(5): 05019005.

Yu, Y., Cai, C. S. He, W. and Peng, H. 2019. Prediction of bridge maximum load effects under growing traffic using non-stationary Bayesian method. *Engineering Structures* 185: 171–183.

Zhao, J., Lin, Z. Tabatabai, H. and Sobolev, K. 2017. Impact of heavy vehicles on the durability of concrete bridge decks. *Journal of Bridge Engineering* 22(10): 06017003.

Zhou, Y. and Chen, S. 2016. Vehicle ride comfort analysis with whole-body vibration on long-span bridges subjected to crosswind. *Journal of Wind Engineering and Industrial Aerodynamics* 155: 126–140.

Zhu, J., Zhang, W. and Wu, M. X. 2018. Coupled dynamic analysis of the vehicle-bridge-wind-wave system. *Journal of Bridge Engineering* 23(8): 04018054.

Zhou, J., Shi, X. Caprani, C. C. and Ruan, X. 2018. Multi-lane factor for bridge traffic load from extreme events of coincident lane load effects. *Structural Safety* 72: 17–29.

11 A Deep Belief Network-based Intelligent Approach for Structural Reliability Evaluation and Its Application to Cable-supported Bridges

Naiwei Lu

Changsha University of Science and Technology, China

Yang Liu

Hunan University of Technology, China

Mohammad Noori

California Polytechnic State University, USA

CONTENTS

11.1 INTRODUCTION

The current transportation market is experiencing rapid growth driven by a steady increase in the global economy, especially in developing countries (Lu et al. 2017). There is an urgent demand on constructing highways and railways in mountainous and cross-sea areas, where long-span bridges are widely built or in construction (Sun et al. 2017). Cable-supported bridges, including cable-stayed bridges and suspension bridges, are widely used in highways crossing gorges, rivers, and gulfs, due to their superior structural mechanical property and beautiful appearance (Sun et al., 2016). A cable-supported bridge is usually the key junction of a highway or a railway demanding a higher safety margin. However, long-span bridges suffer from harsh environmental effects and complex loading conditions, such as heavier traffic loading, more significant wind load effects, severe corrosion effects, and other natural disasters (Shama and Jones 2020; Larsen and Larose 2015; Gong and Agrawal 2016). These effects may subsequently result in changes in the structural mechanical behavior, which may change the dynamic characteristics and resistance of the bridge during its lifetime. Therefore, ensuring the structural safety of cable-supported bridges factoring in a rigorous service environment is a critical task.

A cable-supported bridge has one or more pylons and flexible decks supported by prestressed cables. The cables are critical components to support the girder gravity and moving vehicle loading (Lu et al. 2019a). However, the cable is inclined to corrosion under the coupling effect of vehicular fatigue loading and environmental corrosion. Moreover, cable corrosion will lead to a degradation in strength, which can result in cable rupture or even collapse of the bridge. Research conducted by Mehrabi et al. (2010) indicated that more than half of the stay cables of an in-service cable-stayed bridge were severely corroded and demanded prompt replacement. There are numerous uncertainties in a cable-supported bridge, such as material performance behaviors, cross-sectional dimensions, and various external loadings. These uncertainties make the safety analysis of the cable-supported bridges more complicated. First, structural mechanical behavior has a higher nonlinearity. Second, the bridge structure is statically indeterminate, which leads to a more complicated failure sequence of the system.

The system reliability theory is a powerful tool to estimate the life cycle safety of cable-supported bridges accounting for uncertainties (Jahangiri and Yazdani 2020; Lu et al. 2019b). For a statically indeterminate structure, the scenario of a component failure may not result in safety problems. For instance, the rupture of a stay cable may not affect the safety of the serviceability of the cable-stayed bridge. However, a progressive failure of critical components may result in safety problems or even collapse of the bridge. The system reliability theory can be applied for searching the potential failure sequences and identifying the sensitive components or parameters, which is beneficial for optimization of the design.

There are two critical problems in the structural system reliability evaluation. First and foremost, the long-span bridge has a higher degree of indeterminacy that makes structural performance functions more complicated. Thus, it is important to approximate the limit state function accurately and efficiently by using an adequate approach. In this regard, there are several commonly used approaches including the artificial neural network (ANN) and the response surface method (RSM). With the rapid development of artificial intelligence (AI), the application of an AI-based approach in structural reliability analysis is becoming more popular. Dai et al. (2015) presented a least squares support vector regression (SVR) for structural reliability analysis with a high efficiency and accuracy. Lu et al. (2018) used the SVR approach to estimate the system reliability of a cable-stayed bridge considering the cable damage due to fatigue and corrosion effect. Deep learning technology is a recently developed ANN approach that is widely used in the area of graphics and image recognition (Cui and Li 2019). A deep belief network (DBN) is a type of deep learning technology that is well known for probability training (Wang et al. 2016). The DBN is feasible for approximating the structural mechanical behavior that is the basis for subsequent reliability evaluation.

The second critical problem is how to search the failure sequence of the system efficiently. In this regard, the most commonly used approach is the β-bound approach that uses the range of reliability index to determine the potential failure components. Lee and Song (2012) presented a branch-and-bound method together with the finite element method to update the limit state function. Kang et al. (2008) developed a matrix-based system reliability method to replace the failure tree of complex structures. Liu et al. (2016) developed an adaptive SVR approach combined with the advanced β-bound approach to establish the fault tree of a prestressed concrete cable-stayed bridge.

This study presents a novel and intelligent approach for the system reliability evaluation of cable-supported bridges under stochastic traffic load using deep belief networks (DBNs). Mathematical models for the system reliability of cable-supported bridges were derived with consideration of structural nonlinearities and high order statically indeterminate characteristics. A computational framework is presented to show the procedures for system reliability evaluation using DBNs. In the case study, a prototype suspension bridge was selected to investigate the system reliability under stochastic traffic loading considering site-specific traffic monitoring data. The numerical results indicate the feasibility of using the DBN as a meta-model to estimate the structural failure probability. The dominant failure mode of the suspension bridge was obtained, and the influence of cable degradation due to fatigue-corrosion damage on the system reliability of the suspension bridge was investigated.

11.2 MATHEMATICAL MODEL FOR SYSTEM RELIABILITY OF CABLE-SUPPORTED BRIDGES

11.2.1 Nonlinear Limit State Functions

A cable-supported bridge has long-span flexible girders leading the complexity of structural mechanical behavior. In general, a cable-supported bridge has numerous failure modes. The critical failure modes are the bending moment failure of stiffening girders and cable rupture. The serviceability failure of the structural system includes large deflections and severe vibrations. In this regard, performance functions of the failure modes can be written as

$$Z_i = 1 - \frac{P^i(X)}{P_u^i} - \frac{M^i(X)}{M_u^i} \quad (i = 1, \cdots, m) \tag{11.1}$$

$$Z_j = T_u^{\,j} - T^j\left(X\right) \quad \left(j=1,\cdots,n\right) \tag{11.2}$$

$$Z_u = u_{max} - u\left(X\right) \tag{11.3}$$

where X is the random variables, m and n are the number of potential failure elements of girders and cables, respectively; $P_u^{\,i}$ and $M_u^{\,i}$ are the ultimate axial force and ultimate bending moment for the ith element, respectively; $P^i(X)$ and $M^i(X)$ are the load effect of axial force and bending moment of the pylon and the girder, respectively; u_{max} is the threshold load effect associated with the serviceability of the bridge, u_{max} = $L/500$ for the deflection of bridge girders, where L is the span length of a bridge; and $T_u^{\,j}$ and $T^j(X)$ are the cable tensional strength and the actual cable force, respectively. Note that the limit state functions are nonlinear and implicit, which can be approximated by the following DBNs.

11.2.2 MODELING OF CABLE STRENGTH DEGRADATION

It is acknowledged that the cables in a cable-supported bridge are prone to damage due to corrosion and fatigue damage. The average service time of a bridge cable is usually 20 years. Thus, cable strength degradation should be considered in the evaluation of reliability and safety. In general, cable damage can be considered in a parallel–series relationship model for the purpose of accounting for the effect of the cable length and the number of cables. On the other hand, in order to consider the cable strength degradation due to fatigue-corrosion effect, the cable strength can be derived from the strength of a wire with a Weibull distribution written as (Ma et al. 2020)

$$F_Z\left(z\right) = 1 - \exp\left[-\lambda\left(\frac{z}{u}\right)^k\right] \tag{11.4}$$

where z is the strength of a short wire, and u and k are the distribution parameters in the Weibull function.

In order to model the probability distribution of the parallel–series system, consider a cable consisted of n wires. Faber et al. (2003) provided the probability distribution of the cable based on the aforementioned Weibull distribution. The mean value and the standard derivation are

$$E\left(n\right) = nx_0\left(1 - F_Z\left(x_0\right)\right) + c_n \tag{11.5}$$

$$D\left(n\right) = x_0\left[nF_Z\left(x_0\right)\left(1 - F_Z\left(x_0\right)\right)\right]^{1/2} \tag{11.6}$$

where $c_n = 0.966an^{1/3}$, $a^3 = \dfrac{f_Z^2\left(x_0\right)x_0^4}{\left(2f_Z\left(x_0\right) + x_0 f\left(x_0\right)\right)}$, $x_0 = \left[l\dfrac{L_0}{Lk}\right]^{1/k}\sigma_u$, σ_u is the ultimate stress of a wire, and $f_Z(x_0)$ is a probability density function following the Weibull distribution.

In order to model the cable strength degradation model, consider the experiment study of the strength of corroded cables in the service time of 20 years (Li et al. 2012).

Figure 11.1 shows the probability density function (PDF) of the cable strength of three types of short wires, including a new wire, an uncorroded wire, and a corroded wire. It can be seen that the corroded wire has a lower mean value of strength. In addition, the shape of the distribution totally changed due to the corrosion and fatigue effect.

Based on the practical detection of the research conducted by Lu et al. (2019b), the degradation curves of a cable can be written as

$$y_1(t) = -1.5 \times 10^{-5} t^2 - 3.2 \times 10^{-3} t + 0.998 \tag{11.7}$$

$$y_2(t) = -4.7 \times 10^{-4} t^2 - 2.4 \times 10^{-3} t + 0.996 \tag{11.8}$$

where t is the service time in year, and y_1, and y_2 are the degradation functions for the uncorroded and corroded cables, respectively. It can be concluded that the mean value of the cable strengths for the uncorroded and corroded cables decreased significantly in the service lifetime. In addition, the deviation of the probability was not impacted by the corrosion effect. With a combination of the PDF of the different wires as shown in Figure 11.1 the time-varying PDF of the cable strength can be estimated as shown in Figure 11.2.

As observed from Figure 11.2, the probabilistic characteristics of the corroded and uncorroded cables are totally different. The corroded cable has a higher downturn while the uncorroded cable has a gentle trend of change. This phenomenon demonstrates that corrosion is a critical factor that should be considered in the cable strength modeling. In addition, the degradation model can be utilized for the time-varying reliability analysis of cable-supported bridges. The analytical result could be a theoretical basis for the cable replacement.

11.2.3 Modeling of System Failure

In general, the structural system of a cable-supported bridge can be simplified as several events E_s, which comprises E_i ($i = 1,...,m$). Based on the above assumption, the structural system can be modeled as

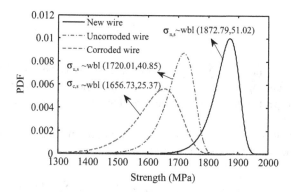

FIGURE 11.1 Comparison of the PDFs of the short wire strength accounting for fatigue damage and corrosion.

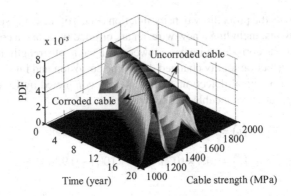

FIGURE 11.2   Degradation of the cable strength accounting for corrosion.

$$
\begin{cases}
E_i = \displaystyle\bigcap_{j=1}^{n} E_i^j \\[2mm]
E_s = \displaystyle\bigcup_{i=1}^{m} E_i
\end{cases}
\tag{11.9}
$$

As shown in Equation (11.9), the system failure events are combined by several failure states in parallel, while a failure state is composed of several components in series. Therefore, the system failure of the bridge can be described in a series–parallel connection system. The critical step is to establish the fault tree of the bridge with consideration of potential failure components. The most commonly used approach to search for potential failure components of a structure is the β-bound method. The principle of the β-bound method will be illustrated below and details of the approach can be found in Liu et al. (2016).

Considering several potential failure components in a system, the critical components should satisfy the reliability range according to the β-bound method:

$$
\beta_{r_k|}^{(k)} = \left[\beta_{\min}^{(k)}, \beta_{\min}^{(k)} + \Delta\beta^{(k)} \right]
\tag{11.10}
$$

where k represents the kth stage of the failure process, $\beta_{r_{kl}}$ is the conditional reliability index, and $\beta_{\min}$ and $\Delta\beta$ are the minimum reliability index and the step of reliability index range, respectively. It is recommended that $\Delta\beta$ is 3 in the first stage and 1 for the following stages.

Once the fault tree is established with the β-bound method, consider the narrow boundary of the structural system, written as

$$
P_{f_{ij}} = \max\left[P_A, P_B \right] + \min\left[P_A, P_B \right] \left(\frac{\pi - 2\arccos\left(\rho_{ij}\right)}{\pi} \right)
\tag{11.11}
$$

$$\begin{cases} P_A = \Phi\left[-\beta_i\right]\Phi\left[-\dfrac{\beta_j - \rho_{ij}\beta_i}{\sqrt{1-\rho_{ij}^2}}\right] \\ P_B = \Phi\left[-\beta_j\right]\Phi\left[-\dfrac{\beta_i - \rho_{ij}\beta_j}{\sqrt{1-\rho_{ij}^2}}\right] \end{cases} \tag{11.12}$$

where β_i and β_j are the reliability index of the ith and jth components, respectively, ρ_{ij} is the correlation coefficient of the ith and jth components, and $\Phi()$ is the cumulative distribution function of the random variable following the standard normal distribution.

11.3 A FRAMEWORK FOR RELIABILITY EVALUATION BASED ON DEEP BELIEF NETWORKS

11.3.1 THEORETICAL BASIS OF DEEP BELIEF NETWORKS

In general, a DBN is a new type of neural network that is usually trained based on the probability principle (Lin et al. 2019). A DBN consists of numerous restricted Boltzmann machines (RBMs), and each RBM is a component of the DBN system. The RBMs in a DBN are connected between hidden layers and visible layers. In the visible layer, the nodes are connected without directions. In the pre-training stage, the unsupervised learning is the main training strategy, where the fine-tuning stage is trained by a supervised learning strategy. The outputs in a visible layer are treated as inputs in the hidden layer. In general, the DBN is widely used in the area of image identification. Its application in structural reliability analysis is relatively insufficient. This study utilized the DBN to approximate the structural response surface functions. The framework of predicting structural load effect using the DNB is shown in Figure 11.3. The principle of the DBN will be illustrated below.

An energy function of a joint structure (V, H) of visible and hidden units is written as

$$E(V,H) = -\sum_{i=1}^{m} a_i v_i - \sum_{j=1}^{n} b_j h_j - \sum_{j=1}^{n}\sum_{i=1}^{m} v_i h_j w_{ij} \tag{11.13}$$

where $V = (v_1, v_2, v_3,\ldots, v_m)$ is a visible layer, $H = (h_1, h_2, h_3,\ldots, h_n)$ is a hidden layer, m and n are the number of elements in the visible and the hidden layer, respectively, $w = (w_{ij})$ is the weight of elements, and a_i and b_j are the bias in the visible layer and the hidden layer, respectively.

The parameters a_i, b_i and w_{ij} are trained and optimized from bottom to top as shown in Figure 11.3 based on an unsupervised learning strategy. Subsequently, the parameters will be adjusted from top to bottom. The training function is

$$w = w + \varsigma\left[P\left(h_j^{(k)} = 1\middle|v^{(k)}\right)\left(v^{(k)}\right)^T - P\left(h_j^{(k+1)} = 1\middle|v^{(k+1)}\right)\left(v^{(k+1)}\right)^T\right] \tag{11.14}$$

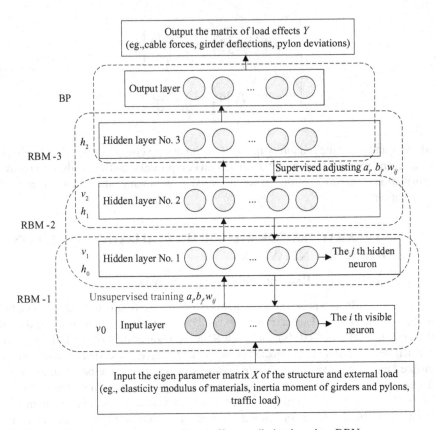

FIGURE 11.3 Flowchart of structural load effect prediction based on DBNs.

$$a = a + \varsigma \left(v^{(k)} - v^{(k+1)} \right) \tag{11.15}$$

$$b = b + \varsigma \left[P(h_j^{(k)} = 1 | v^{(k)}) - P(h_j^{(k+1)} = 1 | v^{(k+1)}) \right] \tag{11.16}$$

where P is a probability function, ξ is the rate of training, which is in the range of 0.05 and 0.2, $h_j^{(k)}$ is the kth optimization of the jth element in the hidden layer.

11.3.2 Proposed Computational Framework

A computational framework shown in Figure 11.4 is presented to illustrate the procedures of utilizing the DBN to estimate the reliability of a cable-supported bridge. Detailed illustrations of the computational procedures are summarized below.

In the first step, select the critical random variables, such as the elasticity modulus of the cable and concrete, and the cross-sectional inertia moment of girders and pylon elements. Training samples should be selected according to the probability distribution of these random variables, which will be utilized for training the DBNs of the bridge structure.

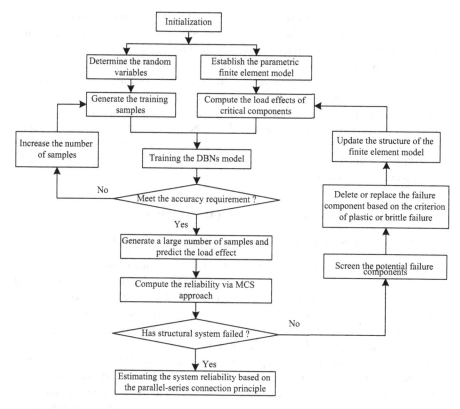

FIGURE 11.4 Flowchart of evaluating system reliability of a cable-stayed bridge via DBNs.

In the second step, establish the parametric finite element model via commercial software, such as ANSYS or MIDAS. It is recommended that ANSYS is used for the finite element modeling, since the APDL code can be connected with MATLAB script files.

In the third step, train the DBNs using the generated samples as the input data and the finite element solution as the output data. The number of hidden layers should be optimized according to the production accuracy of the DBNs. If the accuracy is insufficient, increase the number of training samples, which will be better for the prediction accuracy, but will require more computational effect.

In the fourth step, generate a large number of samples by utilizing MCS according to the probability distribution of the random variables. Predict the load effect of the critical components considering the samples of random variables. Therefore, the structural reliability can be estimated by counting the number of failure samples.

In the fifth step, check the load-carrying capability of the residual structural system considering the potential failure of one of the critical components. The system failure criterion is that the structural system has no load-carrying capacity, or the structural deflection is too large to carry the external load. This limit state is extremely dangerous for in-service bridges since it is possible that the bridge will collapse at any time.

In the sixth step, delete or replace the potential failure components according to the criterion of plastic or brittle failure. For instance, cable failure is brittle, and the corresponding updating approach is to delete the failure cables. The bending failure of a girder is plastic, and thus the corresponding updating approach is to add a plastic hinge at the failure point. Subsequently, the structural system is updated, and the second step returned to for the other circulation computation.

Finally, all of the failure sequences are screened out based on the above procedures. The correlation coefficient matrix should be estimated for the system reliability evaluation. The system event tree can be estimated considering the critical failure sequences. Eventually, the system reliability can be estimated based on the parallel–series connection criterion.

11.4 CASE STUDY OF A SUSPENSION BRIDGE

11.4.1 BACKGROUND OF THE PROTOTYPE SUSPENSION BRIDGE

The prototype bridge is the Nanxi Yangtze River Bridge, which is a suspension bridge on the Yibin-Luzhou highway of China. The general arrangement figure and serial number of each component are shown in Figure 11.5. The main cable is composed of parallel wire strands, the stiffening girders are steel box-girders, and the pylon is concrete box section. The suspenders, the main cable, and the girders are divided into 64 segments according to the construction sequence.

The random variables of the suspension bridge are elasticity modules of the concrete, the steel box-girder and the cable; the cross-sectional area of the main girder, the pylon, and the cable; the unit weights of the steel and the concrete; and the moving vehicle load. Statistical distributions of these parameters are shown in Table 11.1. The upper bound and lower bound for the random variables are determined according to the 3σ principle, where the range of the sampled are determined as "$\mu \pm 3\sigma$."

11.4.2 TRAFFIC LOAD MODELING USING WEIGH-IN-MOTION DATA

The weigh-in-motion (WIM) data collected from a highway bridge in China were selected for probabilistic modeling of traffic loads. Illustration of the WIM system is shown in Figure 11.6, and more detailed information can be found in Lu et al. (2019a). The traffic parameters utilized in the present study are vehicle weights, axle weights, driving lanes, and vehicle spacing. The proportion of trucks and the ratio of truck overloading are 12% and 21%, respectively.

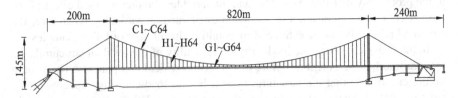

FIGURE 11.5 General arrangement and serial number of components of the Nanxi Yangtze River Bridge.

TABLE 11.1
Statistical Distributions of the Random Variables

Items (unit)	Symbol	Distribution	Mean value	Standard deviation	Upper bound	Lower bound
Elasticity modules of concrete (MPa)	E1	Normal distribution	3.64×10^4	3.64×10^3	2.548×10^4	4.732×10^4
Elasticity modules of steel girders (MPa)	E2	Normal distribution	2.06×10^5	2.06×10^4	1.442×10^5	2.678×10^5
Elasticity modules of suspenders (MPa)	E3	Normal distribution	1.90×10^5	1.90×10^4	1.33×10^5	2.47×10^5
Cross-sectional area of girders (m²)	A1	Lognormal distribution	20.846	1.042	17.72	23.972
Cross-sectional area of the pylon (m²)	A2	Lognormal distribution	26.868	1.343	22.839	30.897
Cross-sectional area of the cable (m²)	A3	Lognormal distribution	1.4×10^{-4}	7.0×10^{-6}	1.19×10^{-4}	1.61×10^{-4}
Unit weight of concrete (kN.m⁻³)	γ1	Normal distribution	26.56	1.33	22.57	30.55
Unit weight of steel (kN.m⁻³)	γ2	Normal distribution	78.5	3.93	66.71	90.29

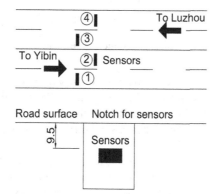

FIGURE 11.6 Weigh-in-motion system of a highway bridge: (a) plane view; and (b) elevation view (unit: mm).

All vehicles were classified into six types based on the vehicle configurations and axle characteristics. The occupancy of different vehicle types and their distribution in each lane are shown in Figure 11.7.

As can be seen in Figure 11.7(a), the light car (V1) has the highest proportion (35.64%), and two-axle (V2) and six-axle trucks (V6) have a higher proportion compared with other types of trucks. As observed in Figure 11.7(b), the light car has a high probability of driving in the fast lane, while heavy trucks mostly drive in the slow lane. This phenomenon is in accordance with the practical rule.

In order to study the probability distribution of axle weights and total weights of heavy trucks, the proportion of axle weight of each axle was assigned as a parameter.

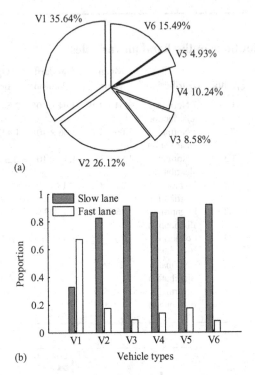

FIGURE 11.7   Proportion of vehicles (a) vehicle types; and (b) driving lanes.

Figure 11.8(a) plots the fitted probability distribution model of V6 trucks, and Figure 11.8(b) shows the relation between the GVW and the axle weight. It is observed that the GVW follows a multimodal distribution, where the truck overloading effect was captured by the probabilistic model. In addition, the axle weight and the GVW are mostly linearly dependent.

Based on estimated probability distribution models, the stochastic traffic flow load model was established using a Monte Carlo simulation. Figure 11.9 plots the simulated dense traffic flow model, where each vehicle was simulated as a label. The stochastic traffic load model contains the parameters of vehicle types, GVW, driving lanes, and vehicle spacing.

11.4.3   Reliability Analysis Based on the DBNs Approach

The finite element model of the bridge was established based on the commercial software ANSYS, as shown in Figure 11.10. The girder element and the pylon element type is the Beam188, and the cable element type is Lin180. The approximated response surface function and training samples of the cable forces are shown in Figure 11.11.

There are three types of failure mode in the bridge components: rupture of a suspender, rupture of the main cable, and bending moment failure of the girder. The reliability index of the bridge components is shown in Figure 11.12.

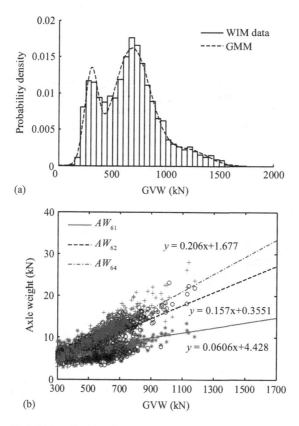

FIGURE 11.8 Probability distribution of V6 trucks: (a) PDF of GVWs; and (b) relation between the GVW and the axle weight.

It is observed that the suspender has a relatively lower reliability index, while the main cable has a higher reliability index. Therefore, the first potential failure component should be the suspenders. For the system reliability evaluation, establishing the fault tree of the structural system is the most important procedure. The fault tree is composed of failure sequences, which can be evaluated based on the β-bound approach.

11.4.4 System Reliability Evaluation

In the first failure searching procedure, the potential failure components are H20 and H44. Failure of the suspender is the brittle failure, and thus the failed cable should be directly removed from the structural system. Subsequently, the structural system is updated and approximated by the DBNs. The potential failure components in the second stage are the stiffening girders associated with G19, G21, G43, and G45. The failure of girders is the plastic failure mode, which should be produced by adding a hinge at the node between two girder elements. At the third failure stage, the main cable is the potential failure component associated with C18, C20, C22, C42, C44,

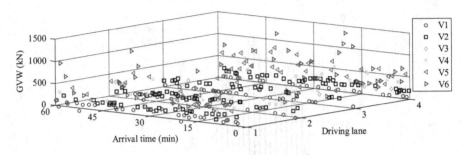

FIGURE 11.9   Stochastic load model for dense traffic flow.

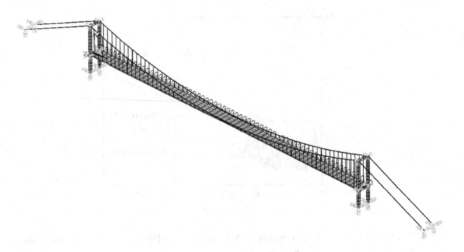

FIGURE 11.10   Finite element model in ANSYS of the suspension bridge.

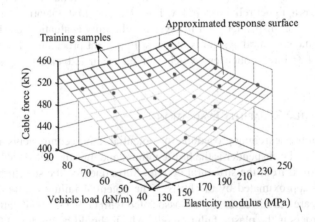

FIGURE 11.11   The simulated response surface and training samples of the cable force based on DBNs.

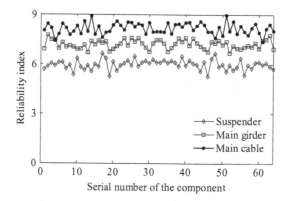

FIGURE 11.12 Reliability indexes of the bridge components.

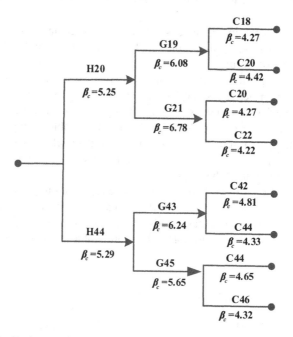

FIGURE 11.13 Fault tree of the suspension bridge.

and C46. The three-level fault tree of the bridge structural system is established as shown in Figure 11.13.

Subsequently, the system reliability of the bridge can be estimated via the series–parallel connection criterion. The corresponding parallel–series relationship is shown in Figure 11.14. It is observed that the system reliability index is 9.63 by considering the three-level system fault tree.

According to the design standard for structural reliability of highway engineering in China, a bridge under threat from brittle failure should have a reliability index greater than 5.2. Thus, the bridge has adequate safety reserve during its lifetime.

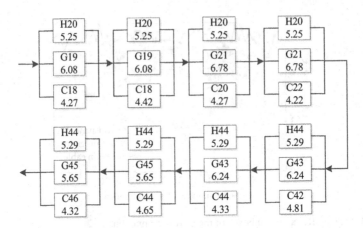

FIGURE 11.14 Series–parallel relationship of the three-level system fault tree.

However, fatigue-corrosion damage is the main factor leading to the degradation of the bridge safety. The strength of cable should be detected and updated for the system reliability evaluation in future.

The cable damage due to fatigue-corrosion effect is the critical factor leading to the shortening of service life, safety threat, and increasing life cycle cost. Thus, influence of the cable damage on the system reliability of the suspension bridge warrants investigation. Considering the degradation model of cable strength as shown in Figure 11.2, the system reliability of the bridge is adaptively updated. Figure 11.15 shows the degradation curve of the system reliability of the suspension bridge in the 30-year service lifetime of the suspender.

It is observed from Figure 11.15 that the cable degradation due to fatigue damage leads the system reliability index to slowly decrease from 9.63 to 7.22. However, the effect of fatigue-corrosion damage leads to the system reliability index decreasing relatively sharply from 9.63 to 4.41. Therefore, cable corrosion is the dominant

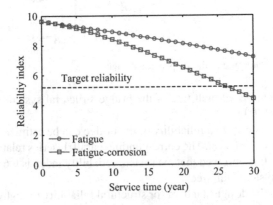

FIGURE 11.15 Influence of the cable strength degradation on the system reliability of the suspension bridge.

factor impacting the bridge safety. In addition, given the target reliability index of 5.2, the suspension bridge has sufficient safety capacity accounting for the cable fatigue damage. Taking into consideration the fatigue-corrosion damage of suspenders, the maintenance time for cable replacement should be 26 years.

11.5 CONCLUSIONS

An intelligent approach is presented for structural system reliability evaluation based on DBNs. Mathematical models for system reliability evaluation of cable-stayed bridges were derived with consideration of the structural nonlinearity and high order statically indeterminate characteristics. A theoretical basis for utilizing DBNs to approximate structural load effect was introduced. A computational framework was presented to show the procedures of evaluating the bridge system reliability via DBNs. The feasibility of the computational framework was demonstrated in a prototype suspension bridge. The conclusions are summarized as follows:

1. The DBN provides an accurate approximation of the response surface of a cable-supported bridge with consideration of the structural nonlinearity and structural system behavior. Thus, as a result of the reliability analysis, DBNs can be utilized as a meta-model for a large number of simulations.
2. The suspender has a relatively lower reliability index, while the main cable has a higher reliability index. Therefore, the first potential failure component is most likely to be the rupture of suspenders. In addition, the degradation of suspenders due to fatigue-corrosion damage has a significant influence on the system reliability of the cable-supported bridge.
3. The system reliability index of the prototype suspension bridge is 9.63 considering the three-level system fault tree. Accounting for fatigue damage and fatigue-corrosion damage, the system reliability index decreases to 7.22 and 4.41, respectively.
4. According to the design standard for structural reliability of highway bridges in China, bridges at risk of brittle failure should have a reliability index greater than 5.2, thus cables should be replaced every 26 years in service time accounting for the fatigue-corrosion damage of suspenders.

It was found that the fatigue-corrosion damage is the main factor leading to the degradation of bridge safety. Therefore, future studies should focus on the cable degradation behavior and its influence on the failure sequence of a bridge. In addition, more inspection data on sites of aged cable bridges is necessary to decide on the cable replacement scheme.

REFERENCES

Cui, T. and Li, S. 2019. Deep learning of system reliability under multi-factor influence based on space fault tree. *Neural Computing and Applications* 31(9): 4761–4776.

Dai, H., Zhang, B. and Wang, W. 2015. A multiwavelet support vector regression method for efficient reliability assessment. *Reliability Engineering and System Safety* 136: 132–139.

Faber, M., Engelund, S. and Rackwitz, R. 2003. Aspects of parallel wire cable reliability. *Structural Safety* 25: 201–225.

Gong, X. and Agrawal, A. 2016. Safety of cable-supported bridges during fire hazards. *Journal of Bridge Engineering* 21(4): 04015082.

Jahangiri, V. and Yazdani, M. 2020. Seismic reliability and limit state risk evaluation of plain concrete arch bridges. *Structure and Infrastructure Engineering* 1–21.

Kang, W., Song, J. and Gardoni P. 2008. Matrix-based system reliability method and applications to bridge networks. *Reliability Engineering and System Safety* 93(11): 1584–1593.

Larsen, A. and Larose, G. 2015. Dynamic wind effects on suspension and cable-stayed bridges. *Journal of Sound and Vibration* 334: 2–28.

Lee, Y. and Song, J. 2012. Finite-element-based system reliability analysis of fatigue-induced sequential failures. *Reliability Engineering and System Safety* 108(12): 131–141.

Li, H., Lan, C. Ju, Y. and Li, D. 2012. Experimental and numerical study of the fatigue properties of corroded parallel wire cables. *Journal Bridge Engineering* 17: 211–220.

Liu, Y., Lu, N. Yin, X. and Noori, M. 2016. An adaptive support vector regression method for structural system reliability assessment and its application to a cable-stayed bridge. *Proceedings of the Institution of Mechanical Engineers, Part O: Journal of Risk and Reliability* 230(2): 204–219.

Lin, K., Pai, P. and Ting, Y. 2019. Deep belief networks with genetic algorithms in forecasting wind speed. *IEEE Access* 7: 99244–99253.

Lu, N., Beer, M. Noori, M. and Liu, Y. 2017. Lifetime deflections of long-span bridges under dynamic and growing traffic loads. *Journal of Bridge Engineering* 22(11): 04017086.

Lu, N., Liu, Y. and Beer, M. 2018. System reliability evaluation of in-service cable-stayed bridges subjected to cable degradation. *Structure and Infrastructure Engineering* 14(11): 1486–1498.

Lu, N., Liu, Y. and Deng, Y. 2019b. Fatigue reliability evaluation of orthotropic steel bridge decks based on site-specific weigh-in-motion measurements. *International Journal of Steel Structures* 19(1): 181–192.

Lu, N., Ma, Y. and Liu, Y. 2019a. Evaluating probabilistic traffic load effects on large bridges using long-term traffic monitoring data. *Sensors* 19(22): 5056.

Ma, Y., Guo Z. Wang L. and Zhang J. 2020. Probabilistic life prediction for reinforced concrete structures subjected to seasonal corrosion-fatigue damage. *ASCE Journal of Structural Engineering* 146(7): 04020117.

Mehrabi, A., Ligozio, C. Ciolko, A. and Wyatt, S. 2010. Evaluation, rehabilitation planning, and stay-cable replacement design for the Hale Boggs bridge in Luling, Louisiana. *Journal of Bridge Engineering* 15: 364–372.

Shama, A. and Jones, M. 2020. Seismic Performance-Based Design of Cable-Supported Bridges: State of Practice in the United States. *Journal of Bridge Engineering* 25(12): 04020101.

Sun, B., Zhang, L. Qin, Y. and Xiao, R. 2016. Economic performance of cable supported bridges. *Structural Engineering and Mechanics* 59(4): 621–652.

Sun, Z., Zou, Z. and Zhang, Y. 2017. Utilization of structural health monitoring in long-span bridges: case studies. *Structural Control and Health Monitoring* 24(10): e1979.

Wang, H., Wang, G. Li, G. Peng, J. and Liu, Y. 2016. Deep belief network based deterministic and probabilistic wind speed forecasting approach. *Applied Energy* 182: 80–93.

Index

Printed in the United States
by Baker & Taylor Publisher Services